IMAGES OF WAR

PATTON TANKS

Rare Photographs from Wartime Archives

MICHAEL GREEN

Pen & Sword
MILITARY

First published in Great Britain in 2012 by
PEN & SWORD MILITARY
an imprint of
Pen & Sword Books Ltd
47 Church Street
Barnsley
South Yorkshire
S70 2AS

ISBN 978 1 84884 761 3

Typeset in Gill Sans by
Phoenix Typesetting, Auldgirth, Dumfriesshire

Printed and bound in England by
CPI Group (UK) Ltd, Croydon, CR0 4YY

Pen & Sword Books Ltd incorporates the Imprints of Pen & Sword Aviation, Pen & Sword
Family History, Pen & Sword Maritime, Pen & Sword Military, Pen & Sword Discovery,
Wharncliffe Local History, Wharncliffe True Crime, Wharncliffe Transport, Pen & Sword
Select, Pen & Sword Military Classics, Leo Cooper, The Praetorian Press, Remember When,
Seaforth Publishing and Frontline Publishing

For a complete list of Pen & Sword titles please contact
PEN & SWORD BOOKS LIMITED
47 Church Street, Barnsley, South Yorkshire, S70 2AS, England
E-mail: enquiries@pen-and-sword.co.uk
Website: www.pen-and-sword.co.uk

Contents

Acknowledgments

Besides those individuals who contributed pictures to the book and are credited in the captions, the author would like to thank a number of other individuals for their help in the completion of this book. These include Michael Panchyshyn, Randy Talbot, Don Moriarty, Lee Burnell, Phil Hatcher, Michael Brandt, Todd Armstrong, James Brown and Jim Mesko.

Institutions that assisted the author include the Military Vehicle Technology Foundation, the now closed Patton Museum of Armor and Cavalry, and TACOM (Life Cycle Management Command). Pictures credited to the former Patton Museum of Armor and Cavalry are shortened to just the Patton Museum in the name of brevity. Pictures credited to TACOM (Life Cycle Management Command) are shortened to TACOM for the sake of brevity.

Dedication

To my friend and mentor the late Richard P. Hunnicutt

Foreword

Aconcise book on the Patton series medium tanks has long been needed. This comes as a surprise for many of us, given the tank's long service in various versions and in so many of the armies of the world. But authors have too often skipped over it as perhaps too common or ordinary. Indeed Patton series tanks became ubiquitous in NATO and other allied armies of the 1950s and 1960s, largely because of their voluminous production in the United States and the equally rapid succession of improved variants.

Accelerated in development and production by the onset of the Korean War of 1950-53, the Patton series tanks reflected all the strengths and weaknesses of the American tank research and acquisitions system and the supporting industry. The resulting combinations of new and unproven components with otherwise sound materiel illustrated best that the complexity of tank designs and development rivaled that of warships and combat aircraft.

Despite their disappointing fuel consumption and short radius of action, the Patton series tanks of the 1950s demonstrated increasing improvements in ease of handling, crew comfort and simplicity of operation. The U.S. cross-drive transmissions made driving a sheer joy, and the simple yet effective mechanical-analog fire control system gave excellent results with the proven and well-liked 90mm cannon. By the time the M48 fleet was rebuilt to the M48A3 model in the early 1960s one could say that U.S. tank design and development had at last matured.

I climbed into my first of these in early 1970 after only eight months as a Marine Corps officer. With its diesel engine, coincidence rangefinder and NBC gas particulate system, the tank now represented the best that 1960s modern armor was intended to be. We trained into the tank easily and soon were able to direct other new personnel joining our units. The M48A3 was robust, forgiving and totally reliable over long periods of operation. As heavy as it was, the power pack took it easily over most terrain from high surf beaches to boulder-strewn mountains that we encountered in the United States and a Mediterranean deployment.

Driving our M48A3 tanks became so second nature that one had to restrain reaching for the nonexistent turn signal handle because it drove so much like an automobile, so smooth was its steering, braking and automatic shifting. Only in reverse did the steering cause a surprise or two. The gun was just a sweet shooter, with a variety of ammunition never again seen in tanks and, with very few shots in training, a crew could score first round hits at ease beyond 2000 meters. I had to sympathize with the old salts who received the news badly in 1971 that our guns were no longer considered effective against first line tanks and we would change to the 105mm and the M60 series, which we never found so user friendly.

Patton tanks served long and mostly well, and we enjoyed operating them and respected their combat record. Nostalgia for them remained strong even as we moved into the M60 and M1 series that finished the century.

Kenneth W. Estes
Lieutenant Colonel, U.S. Marines
Author, *Marines Under Armor* (2000)

Chapter One:

Patton Tank Genesis

Even as the first versions of the American-designed and built M4 series medium tanks entered into production in February 1942, the U.S. Army became aware that German tank development showed a trend towards greater levels of firepower and armor protection levels. To address these issues, the U.S. Army began a development program in the spring of 1942 aimed at

The U.S. Army's answer to late-war German Panther and Tiger tanks was the M26 heavy tank. Unlike the thinly armored and under-gunned M4 series medium tanks, the new tank boasted fairly thick armor and a high-performance, tank-killing M3 90mm main gun. The restored vehicle pictured belongs to a private collector. *Bob Fleming*

Pictured is a U.S. Army M26 heavy tank belonging to the 9th Armored Division moving along the road between *Thum* and *Ginnick*, Germany, on March 1, 1945. The M26 tank was officially nick-named the "General Pershing" after the Second World War and at the same time reclassified as a medium tank. *Patton Museum*

developing a replacement for the M4 series of tanks, officially nicknamed the "Sherman" by the British Army. Funding would not be a problem, as the purse strings of the U.S. Congress had loosened following the Japanese attack on the American naval base in Hawaii in December 1941. This marked America's official entry into the Second World War.

The U.S. Army realized early on that the successor to the M4 series tanks had to have superior firepower, armor protection, and mobility to deal with future German tank developments. As a starting point, the U.S. Army decided to incorporate the lessons learned in the battles of North Africa as well as all the technical advances that had taken place since the design of the first versions of the M4 series tanks had been finalized. After a great deal of resistance from many within the U.S. Army who felt that nothing more than an up-gunned M4 series tank would suffice to deal with future German tank developments, the much-delayed end result of this developmental work was the series production of the

92,355lbs (42mt) T26E3 tank beginning in November 1944. The prefix "T" following by numerals meant that a vehicle was an experimental model still subject to modifications.

Unlike the under-gunned early production 66,800lbs (30mt) combat loaded M4 series tanks armed with a 75mm main gun and the 74,200lbs (34mt) combat loaded late production Sherman tanks armed with a 76mm main gun, the new five-man T26E3 tank boasted a long and powerful 90mm main gun capable of killing late-war German Panther medium tanks and Tiger heavy tanks. The 90mm main gun mounted on the T26E3 tank was designated the M3.

Like the main guns mounted on the late-war German tanks, the M3 90mm main gun on the T26E3 tank featured a muzzle brake to help reduce recoil and dust. Unlike the thinly armored M4 series tanks, the armor protection levels on the T26E3 tank were fairly robust, surpassing that on the German Panther or Tiger tanks in some regards.

The first U.S. Army T26E3 tank knocked out in combat during the Second World War was serial number 38 and nicknamed "Fireball" by its crew. It took three hits from an 8.8cm gun mounted on a German Army Tiger E heavy tank in late February 1945. Pictured is the third strike on the vehicle that gouged out a chunk of the turret. *Patton Museum*

Due to an overwhelming desire by U.S. Army tankers fighting in the European Theater of Operations (ETO) to have a vehicle in service that was at least an equal in fighting effectiveness to the Panthers and Tigers, twenty T26E3 tanks out of the initial production run of forty vehicles were rushed to Western Europe, arriving at the Belgium port of Antwerp, in January 1945. The twenty tanks were part of technical mission, code named "Zebra," intended to assist in the rapid introduction of the new tank as well as several other weapons to the ETO. To evaluate how the twenty T26E3 tanks would fare in combat against the German Panther or Tiger tanks, the U.S. Army 3rd and 9th Armored Divisions each received ten vehicles.

The first tank on tank combat action took place on 26 February 1945, when a T26E3 tank from the 3rd Armored Division, nicknamed "Fireball," was knocked out by German Tiger E heavy tank. The following day another T26E3 tank from the same division avenged Fireball by destroying a German Tiger E heavy tank and two Panzer IV medium tanks.

On March 1, 1945, another U.S. Army T26E3 tank (serial number 22) was knocked out by two German high explosive (HE) rounds that caused extensive damage to the vehicle's turret and running gear as seen in this picture. Notice the tank commander's cupola on the ground next to the vehicle. *Patton Museum*

Here we see the first U.S. Army M26 tanks assigned to the 11th Armored Division in early April 1945. At the time, the division was assigned to General George S. Patton's famous Third Army. Due to the lateness of their arrival there would be no combat engagements between Third Army M26 heavy tanks and their German late-war counterparts. The chalk marks on the glacis indicate the engine oil was changed on 14 April 1945. *Patton Museum*

In March 1945, the T26E3 tank was re-designated as the M26 heavy tank. The prefix "M" followed by numerals meant that a vehicle design was considered "type classified" and accepted into the American military inventory. The prefix "E" following a vehicle designation meant that an experimental modification was made to the vehicle, and E1, E2 and so forth indicates additional experimental modifications added to the manufacturing of a vehicle but not constituting a new model number. Upon type classification of a vehicle, the E modification is incorporated into the tank's final design.

During the fighting for the German city of Cologne on 6 March 1945, an M26 heavy tank of the 3rd Armored Division destroyed a German Panther medium tank with three shots. This "tank versus tank" incident was captured on film by a U.S. Army Signal Corps cameraman and is often seen on television shows about the war in Europe.

The best known combat action in which the M26 tanks of Operation Zebra took part, but did not include any tank-versus-tank action, occurred on 7 March

On the production line at the Fisher Tank Arsenal during the Second World War is a long line of M26 tanks. Production of what was then designated the T26E3 tank began at Fisher in November 1944. The initial order called for 250 T26E3 tanks to be built. *Patton Museum*

A single pilot example of the T26E4 tank was sent to Germany just before the war in Europe ended. It was armed with a new more powerful T15E1 90mm main gun. To compensate for the gun's weight and length, two cylinders containing coil springs were mounted on the vehicle's turret as seen in this picture. To increase its armor protection levels, applique armor was mounted on the front hull and turret of the vehicle by a maintenance battalion in Europe. The T26E4 is now commonly referred to as the "Super Pershing." *Patton Museum*

1945 when four M26 tanks of the 9th Armored Division aided in the capture of the *Ludendorf* railroad bridge over the Rhine River at the German town of *Remagen*.

By the end of March 1945, forty additional M26 tanks would arrive in Western Europe. These would be assigned to the American Ninth Army, with twenty-two going to the 2nd Armored Division and eighteen to the 5th Armored Division. In early April 1945, the 11th Armored Division of General George S. Patton's Third Army would receive thirty M26 tanks. However with the war in Europe winding down, there were no additional tank-versus tank-combat actions between M26 tanks and German Tiger or Panther tanks. By VE-Day (Victory in Europe Day) on 8 May 1945, there were 310 M26 tanks in Europe, with 200 of them issued to field units.

With the war in Europe ending, greater attention was paid to the fighting in the Pacific against the Empire of Japan. On the island of Okinawa, Japanese 47mm towed antitank guns were taking a heavy toll of thinly armored Sherman tanks. The perceived solution to this problem was the dispatch of twelve M26 tanks to that theater of operation. The M26 tanks shipped to Okinawa arrived on the island after fighting concluded on 21 July 1945. The M26 tanks were then envisioned as playing an important role in the planned follow-up invasion of Japan; however, the Japanese surrender on V-J Day (Victory over Japan Day) on 15 August 1945, brought a quick end to that mission.

The M26 tank was officially designated the "General Pershing" after World War II, with most simply referring to it as the "M26" or the "twenty-six." In May 1946, the M26 tank was reclassified as a medium tank as the U.S. Army was envisioning developing much larger and heavier tanks.

The next combat action for the M26 tank would occur during the Korean War, which began with the North Korean invasion of South Korea on 25 June 1950. The North Korean Army invasion of its southern neighbor was spearheaded by 150 T34/85 medium tanks supplied by the Soviet Union. As the American-equipped South Korean Army had no tanks, and the U.S. Army divisions based in the Pacific had only M24 "Chaffee" light tanks—no match for the Soviet-built tanks—a pressing need quickly developed to provide the American

The front hull of the M26 tank consisted of a large cast homogenous armor (CHA) piece. Ventilation for the vehicle's interior was provided by a blower located on the hull roof between the drivers. It was the blower and its intake that caused the large bulge in the top center of the front hull roof that is very evident in this picture. *Patton Museum*

Belonging to the former Patton Museum of Armor and Cavalry is this unrestored M26 tank that came off a U.S. Air Force firing range. A distinguishing external feature of the M26 was the large and squat CHA turret that set well forward on the hull as seen in this photograph. *Michael Green*

ground forces in theater with armored fighting vehicles that could kill a T34/85 tank. The solution was the shipment of a late production version of the M4 series tanks armed with a 76mm main gun, and M26 tanks to South Korea.

The first three M26 tanks arrived in South Korea in July 1950. However, due to their poor mechanical condition they promptly broke down and were abandoned. The next batch of M26 tanks began showing up in early August 1950. The first tank-versus-tank combat action between M26 tanks and North Korean T34/85s occurred just before dark on 17 August 1950. Three U.S. Marine Corps M26 tanks took part in an engagement with four North Korean T34/85 tanks, with the American tank crews claiming the destruction of three of the enemy tanks. For the next six months, U.S. Army and Marine Corps M26 tanks would see heavy action against the North Korean Army. The M3 90mm main gun on the M26 tank had no problem penetrating the armor on the T34/85 tank, and in conjunction with superior optical sights would allow the American tankers to normally prevail in any tank-versus-tank action.

An upgraded M26 tank designated the M26A1 tank was introduced into

The crewmen of a postwar U.S. Army M26 tank are shown loading their vehicle with main gun ammunition. There was storage space in the M26 tank for 70 main gun rounds. The running gear of the vehicle consisted of 12 individually sprung dual road wheels (six per track) with 10 dual track return rollers (five per track). The tank rode on a torsion bar suspension system. *Patton Museum*

service during the Korean War. The prefix "A" following a vehicle designation indicates a substantial change to the design, such as the fitting of a new engine, main gun or so forth; continuations such as A1, A2, A3 identify more standardized modifications.

The differences between the M26A1 tank and the original M26 tank included the fitting of an improved 90mm main gun, designated the M3A1, with a lighter weight single baffle muzzle brake. The lighter weight muzzle brake meant a lighter equilibrator spring could be used in the gun mount to balance the main gun. The new M3A1 90mm main gun on the M26A1 tank also featured a bore evacuator fitted just behind the new muzzle brake. The bore evacuator was a Second World War British invention that helped to eliminate powder fumes from the turret when the main gun was fired.

When the frontlines between the American ground forces fighting under the banner of the United Nations, and the North Korean ground forces, moved into the more mountainous terrain of Korea, the M26 series tanks quickly fell out of favor with its crews. The reason for this was not the tank's firepower or armor protection, but its 500-horsepower gasoline engine, the same as mounted in the late production versions of the M4 series tanks it served alongside in the conflict. Adequate for the much lighter M4 series tanks, the same engine could not muster the necessary horsepower to push the heavier bulk of the M26 series

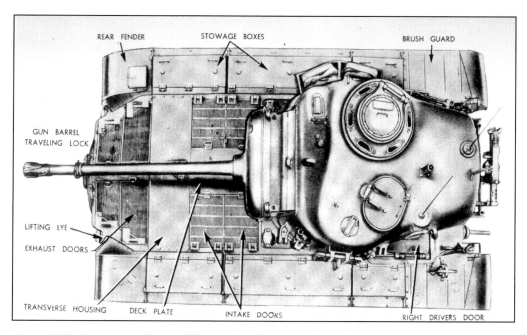

From a U.S. Army manual comes this overhead picture of an M26 tank with its various external features identified. With the main gun forward, the vehicle was 28 feet (8.6m) long and with sand shields fitted it had a width of 11 feet 6 inches (3.5m). To the top of the cupola the vehicle was 9 feet 1.40 inches (2.8m) tall. *Patton Museum*

tanks up the same Korean hills. The fact that the M26 series tanks were under-powered was recognized early on in its development and was being addressed as early as 1943 with new more powerful tank engines under development. Sadly, the next generation of more powerful engines for American tanks did not appear until 1946.

The power pack for the M26 tank consisted of the Ford liquid-cooled GAF gasoline engine, a torqmatic transmission, and a controlled differential. All of these components could be removed from the tank as a single assembly, pictured here. *Patton Museum*

This view of an M26 tank shows the engine exhaust muffler on the upper rear portion of the rear engine compartment plate. Notice the turnbuckles and rods holding up the rounded rear fender extensions on this vehicle, which was a postwar addition to the vehicle. These turnbuckles also appeared on the front of the vehicle. *TACOM*

FINAL DRIVE LEVEL AND FILL PLUG

FINAL DRIVE DRAIN PLUG

ENGINE COMPARTMENT LEFT REAR DRAIN VALVE

ENGINE CRANKCASE DRAIN PLUG COVER

FUEL TANK DRAIN PLUG COVERS

LEFT REAR HULL DRAIN VALVE

DIFFERENTIAL DRAIN PLUG COVER

TRANSMISSION DRAIN PLUG COVER

RADIATOR DRAIN PLUG COVER

AUXILIARY ENGINE CRANKCASE DRAIN PLUG COVER

ENGINE COMPARTMENT RIGHT FRONT DRAIN VALVE

WHEEL ARM SUPPORT

TORSION BAR SPRING ANCHOR PLUG COVER

RIGHT FRONT HULL DRAIN VALVE

ESCAPE HATCHES

From an U.S. Army manual comes this photograph of the hull bottom of an M26 tank (looking forward from the rear of the tank) with all the various components listed. The two escape hatches at the front of the hull were seen as weak spots by crewmen because an antitank mine blast would often push them into the hull interior. *Patton Museum*

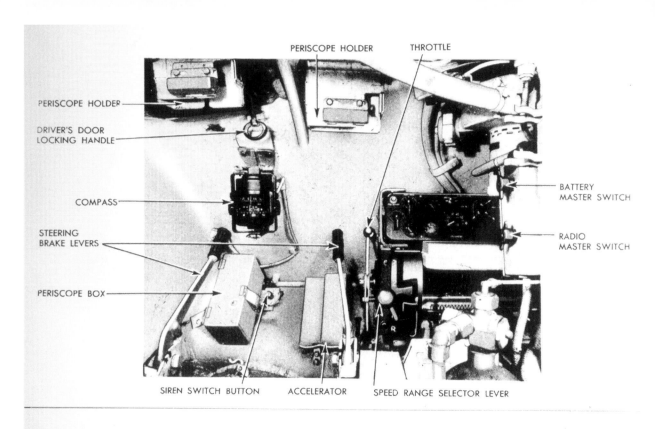

PERISCOPE HOLDER

THROTTLE

PERISCOPE HOLDER

DRIVER'S DOOR
LOCKING HANDLE

BATTERY
MASTER SWITCH

COMPASS

RADIO
MASTER SWITCH

STEERING
BRAKE LEVERS

PERISCOPE BOX

SIREN SWITCH BUTTON   ACCELERATOR   SPEED RANGE SELECTOR LEVER

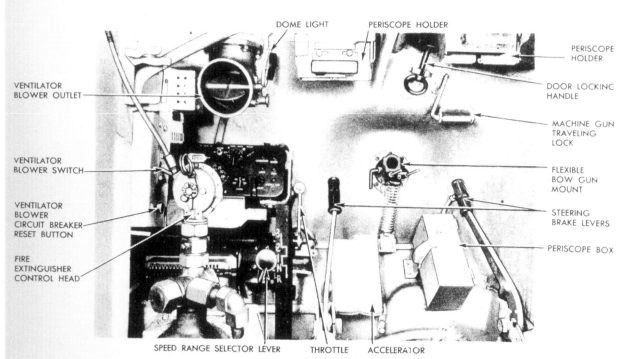

DOME LIGHT   PERISCOPE HOLDER

PERISCOPE
HOLDER

VENTILATOR
BLOWER OUTLET

DOOR LOCKING
HANDLE

MACHINE GUN
TRAVELING
LOCK

VENTILATOR
BLOWER SWITCH

FLEXIBLE
BOW GUN
MOUNT

VENTILATOR
BLOWER
CIRCUIT BREAKER
RESET BUTTON

STEERING
BRAKE LEVERS

FIRE
EXTINGUISHER
CONTROL HEAD

PERISCOPE BOX

SPEED RANGE SELECTOR LEVER   THROTTLE   ACCELERATOR

In these pictures from an U.S. Army manual appear the driver's position on an M26 tank on the left and the assistant driver/bow gunner's position on the right. The bow .30 caliber (7.62mm) machine gun is not mounted in this picture. *Patton Museum*

90-MM STOWAGE COMPARTMENT COVERS

COLLECTOR RING BOX

WIRING HARNESS

90-MM SIDE STOWAGE RACK

FIGHTING COMPARTMENT REAR FLOOR PLATES

The bulk of the main gun rounds for the 90mm cannon mounted in the M26 tank were located in the bottom of the hull in various storage lockers just below the vehicle's turret basket as seen in this picture from an U.S. Army manual. To access these main gun rounds the turret would have to be traversed in different directions. *Patton Museum*

The M26 tank would see extensive action during the Korean War in a wide variety of roles, ranging from tank-on-tank combat to infantry support duties. Pictured is a U.S. Army M26 tank on the Korean battlefield. Notice the machine gun ammunition containers stored on the vehicle's right fender. These machine gun ammunition containers were typically stored on the horizontal sections of both the vehicle's right and left fenders. *Patton Museum*

An M26 tank belonging to the now closed Patton Museum of Armor and Cavalry is shown being prepared for shipment to Fort Benning, Georgia, where it form part of the vehicle collection of the future National Armor and Cavalry Museum. *Chun-Lun-Hsu*

The only distinguishing external features on the M26A1 tank were the revised single baffle muzzle brake in place of the double baffle muzzle brake on the M26 tank and the addition of a bore evacuator behind the muzzle brake as seen here in this picture of the M3A1 90mm main gun on an M26A1 tank. *Chris Hughes*

Looking down through the tank commander's cupola on an M26A1 tank, the breech of the M3A1 90mm main gun as well as the loader's ready rounds are visible. The M3A1 90mm main gun in the M26A1 tank was an improved version of the M3 90mm main gun mounted in the M26 tank. *Michael Green*

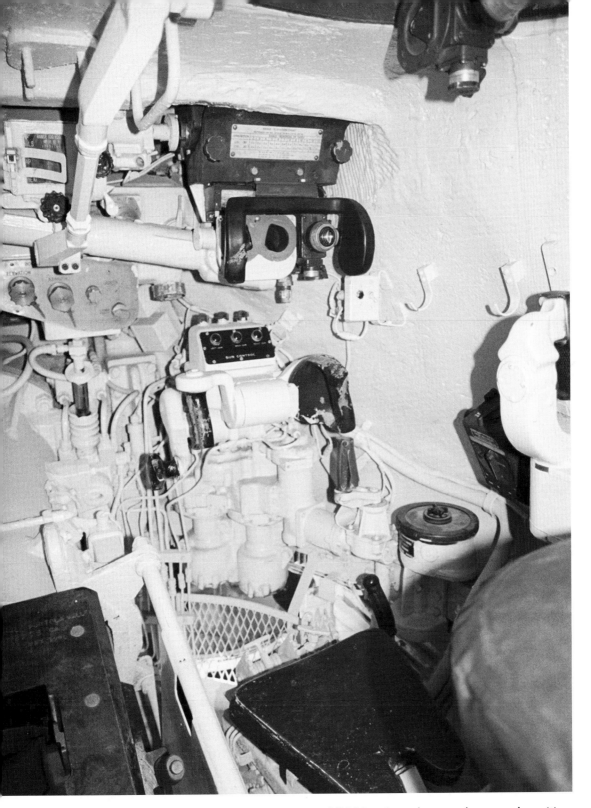

From just behind the M3A1 90mm main gun in a M26A1 tank can be seen the gunner's position. A portion of the tank commander's seat can be seen in the right foreground of the picture. *Michael Green*

Chapter Two

M46 Patton Tank

The early postwar American monopoly on the atomic bomb led its senior political and military leadership to believe the threat of another large-scale ground war had greatly receded. In line with this, funding devoted to the development of the next generation of ground weapons for the United States armed forces, such as tanks, was slashed. With no money to replace the Second World War era M26 series tanks, the U.S. Army decided in January 1948 as an interim measure to modernize its inventory of roughly 2,200 M26 series tanks.

The biggest design shortcoming of the M26 series tanks had always been the relatively low power of their liquid-cooled Ford GAF gasoline engines that

The financial constraints that forced the US to modernise rather than replace its M26 stocks led to the construction of ten vehicles designated T40, one of which is shown here. *Patton Museum*

A T40 tank is shown here at the U.S. Army Aberdeen Proving Ground on October 3, 1949. The engine exhaust on the T40 tank was ejected through the top center grill out pipes extending sideways to mufflers mounted on each fender, as seen in this picture. *Patton Museum*

produced only 500 horsepower. Fortunately, the U.S. Army had initiated the development of the "ideal" tank engine in July 1943. What eventually sprung forth from this line of development was the Continental Motors Corporation AV-1790-1 gasoline engine, which boasted 740 gross horsepower. It was a 12-cylinder, V-type, four-cycle, air-cooled engine.

In 1946, one of the first three examples of the new, more powerful Continental AV-1790-1 engine was installed in a modified M26 series tank that was then designated as the M26E2 tank. Coupled to the new engine was a newly-designed Allison Corporation CD-850-1 cross-drive transmission, which transmitted the power generated by the engine to the tank's final drives and drive sprockets. The CD-850-1 was referred to as a cross-drive transmission because of its transverse mounting in the rear hull engine compartment of the M26E2 tank.

After testing at the Detroit Tank Arsenal, the single M26E2 tank was shipped to Aberdeen Proving Ground, Maryland, home of the U.S. Army's Ordnance

Branch. Positive results with the M26E2 tank led to the authorization of ten more to be designated as the T40 medium tanks. The T40 tanks had a new power pack arrangement consisting of an upgraded Continental AV-1790-1 engine designated as the AV-1790-3, which boasted 810 gross horsepower. It was coupled to an improved version of the original Allison Corporation CD-850-1 cross-drive transmission, designated CD-850-5.

The new power pack arrangement required some design changes to the top and rear of the T40 tank hull. The most noticeable external changes on the T40 tank were the engine exhaust pipes that extended out sideways from the roof of the tank's engine compartment to mufflers mounted on each of the vehicle's rear fenders. There were also three square armored access hatches on the lower vertical rear face of the tank's rear hull plate for servicing of the CD-850-5 cross-drive transmission.

Another external change to the T40 tank was the addition of a small track tension idler between the dual rear road wheels and the rear hull mounted drive

With the rear engine compartment plate free of the engine exhaust muffler seen on the M26 series tanks, the U.S. Army had three square armored covered access ports for the cross-drive transmission installed on the T40 tank, as seen in this picture. Just above the access ports is the large external interphone control box. *Patton Museum*

=DETROIT ARSENAL=
NEG. NO. 21335   DATE 21 March 1950   DEVELOPMENT & ENGINEERING
Power Package for Tank, Medium, M46. 3/4 Right Front.

The big advantage the T40 tank had over the M26 series tanks was the new smaller, more powerful, and compact air-cooled Continental gasoline-powered AV-1790-1 engine, combined with the new General Motors CD-850-1 cross-drive transmission as seen here. *TACOM*

sprockets on either side of the vehicle's suspension system. This was done to prevent the tank's tracks being shed on sharp turns or when traveling over rough terrain. Like the front compensating idler (which appeared on the entire Patton tank series) it also helped to eliminate slack in the track under dynamic conditions such as hard braking. The track tension idler was originally referred to as the compensating idler wheel.

The U.S. Army considered mounting a new, more powerful 90mm gun on the T40 tank. In the end, a decision was made to use a modified M3 90mm gun, designated the M3A1, on the T40 tank. Unlike the M3 90mm main gun, the

U.S. Army pleasure with the testing of the T40 tank led to the decision on July 30, 1949 to standardize the vehicle as the medium tank M46. A feature seen on the M46 tank pictured and not on the M26 series tank is the tension idler wheel located behind the last set of road wheels and the rear-mounted drive sprocket, which is mounted higher than the rear-mounted drive sprocket on the M26 tank. *TACOM*

M3A1 90mm main gun sported a bore evacuator as well as a new, lighter and smaller single baffle muzzle brake. The optical sighting system on the T40 tank was improved with the addition of a new M83 sighting telescope.

Pleased with the results of testing the T40 tank, the U.S. Army decided on 30 July 1948 to standardize the vehicle as the medium tank M46. The vehicle also received the official name the "General Patton" in honor of the late General George S. Patton of Second World War fame. Most American tankers simply referred to it as the "forty-six."

According to the U.S. Army's TACOM (Life Cycle Management Command) historical office, between 1948 and 1951, a total of 1,170 M26 series tanks were converted by the Detroit Tank Plant to the M46 tank configuration.

The last 360 production units of the M46 tank featured a number of modifications based on continued testing and user input from the field and were assigned the designation medium tank M46A1. The original plan calling for converting almost all 2,200 M26 tanks in the U.S. Army's inventory into the M46 tank configuration proved impossible to complete as some of the M26 tanks had been diverted to take part in the Korean War.

The new M46 series tank differed externally from the T40s tanks that preceded them by having three circular armored access hatches on the lower vertical rear face of the vehicle's rear hull plate, rather than the three square armored access hatches seen on its predecessor. Instead of the Continental AV-1790-3 engine as fitted to the T40 tank, the M46 series tanks would receive power from progressively improved models of the same engine, the last one bearing the designation AV-1790-5B. The M46 series tanks weighed in at about 97,000lbs (44mt) combat loaded.

Like the M26 series tanks, the M46 series tanks were rushed to South Korea to help stem the North Korean onslaught, the first ones arriving in August 1950 as part of the U.S. Army's 6th Tank Battalion. Eventually 200 M46 series tanks

RA PD 122679

Figure 243. Left rear road wheel arm and compensating idler wheel.

The arrangement of the tension idler wheel on the T40 and the M46 tanks is seen here in this picture from a U.S. Army manual. The track tension idler wheel was sprung by a small torsion bar and was intended to maintain track tension during turns and when operating over rough terrain thus reducing the chances of the tank's tracks coming off. The U.S. Army originally referred to the tension idler as the compensating idler wheel, which is noted in this picture. *Patton Museum*

As with the T40 tank, the M46 tank retained the turret and forward hull structure of the M26 series tanks as is clearly visible in this picture of an M46 tank. Notice the various storage boxes on the vehicle's horizontal fender sections. Both the tank commander's and loader's overhead hatches could be locked from inside. *TACOM*

would be deployed to Korea. They would serve alongside roughly 300 M26 series tanks and 680 later production versions of the M4 series tanks, the majority armed with a 76mm main gun.

Of the three types of American medium tanks to see service during the Korean War, the most favored for tank-versus-tank combat were the 90mm

A picture from the production line shows welding being done on the hulls of M46 tanks. The hull of the M46 tank was 20 feet 10 inches (6.36m) long and retained the same armor thickness as the M26 tank series. *TACOM*

main gun equipped M26 series tanks and the M46 series tank. Tank-versus-tank action in Korea, however, became extremely rare after November 1950.

For American tanks, the greatest cause of losses in the Korean conflict was mechanical failure, with the engines in the M26 and M46 series tanks being the biggest culprit. The next most common mechanical failure with the same tanks where their transmissions, including the clutches and gearing; with the M46 series tank transmission being the most trouble prone. Neither tank was up to the demands imposed by the mountainous and heavily-wooded Korean terrain. The biggest enemy-caused casualty producer for American tanks during the Korean conflict was antitank mines.

Like the other American medium tanks to see action during the Korean War, the main role of the M46 series tank generally became infantry support, with occasional stints as artillery pieces when poor terrain conditions or weather ruled out the use of the vehicle in its normal roles. Other roles for the M46 series tanks included bunker busting, which the U.S. Army eventually considered as being extremely uneconomical and wasteful in main gun ammunition expenditure for the minor results achieved. The British experience of bunker-busting with tanks, however, was exactly the opposite, with the British Centurion earning a great reputation for this work. There were also armored raids and

Belonging to the former Patton Museum of Armor and Cavalry is this early-production M46 tank, which is minus the sheet metal muffler shields. The turret of the M46 tank was traversed either manually or with an electric-hydraulic system. With the powered traverse system the turret on the M46 tank could be turned 360 degrees in 15 seconds. *Dean and Nancy Kleffman*

reconnaissance in force operations conducted during the Korean War which achieved excellent results against their Communist opponents.

A U.S. Army research report titled *"Employment of Armor in Korea: The Second Year,"* issued in April 1953, concluded that the later production versions of the Sherman tanks that served during the conflict were more mechanically reliable than the M26 and M46 series tanks and easier to maintain. Some of the blame for the poor mechanically reliability for these tanks can be attributed to the general failure of the U.S. Army logistical system to push the needed spare parts to the tanks in the frontlines. Added to this problem were the lack of training and experience among the tank mechanics tasked with keeping the M46 series tanks running, as well as the poor level of training among the men assigned to the vehicle. To make matters even worse, the weight of the M46 series tanks overtaxed the M4 series tank-based armored recovery vehicles employed in Korea.

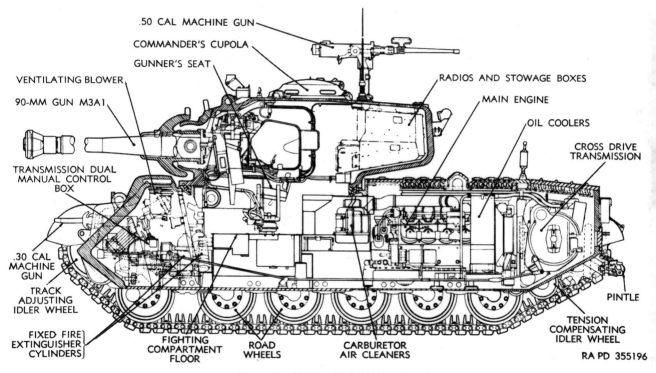

Figure 7. Medium tank M46, sectionalized.

A sectional view of the M46 tank comes from a U.S. Army manual. The overall thickness and slope of the M46's armor layout is illustrated. Fire protection for the vehicle came from three fixed 10 pound carbon dioxide fire extinguishers in the front hull, which are identified in this sectional view. There were also two portable four pound carbon dioxide fire extinguishers carried inside the vehicle. *Patton Museum*

The demands of the Korean War soon resulted in a number of M46 tanks being sent off to South Korea. Here we see an M46 being unloaded from the bowels of a merchant ship. Notice the early version of the sheet metal cover over the engine exhaust muffler. The M46 tank carried 232 gallons (878L) of 80 octane gasoline. *Patton Museum*

The M46 tank was armed with the M3A1 90mm main gun, the same as mounted on the M26A1 tank. Distinguishing features of the M3A1 were the single baffle muzzle brake and bore evacuator located just behind the muzzle brake as seen on the vehicle pictured. The maximum firing rate of the main gun was eight rounds per minute. *TACOM*

Coming off the ramp of a U.S. Navy Landing Ship Tank (LST) is a U.S. Army M46 tank nicknamed "Fighting Fool." The crew of the vehicle added a makeshift storage bustle at the rear of the turret to carry more of the things they need for long stretches in the field. There was storage space for 70 main gun rounds in the M46 series tanks. *Patton Museum*

Marines are shown installing heavy wire fencing material around the turret of a M46 tank, sometime during the Korean War. The wire screen was intended to protect the tank from the high explosive antitank (HEAT) warheads of enemy rocket launchers. Notice that the M46 tank has three circular cross-drive transmission access panels instead of the three square ones on the T40 tank. *Patton Museum*

The biggest problem during the Korean War with the M46 tank was keeping them working properly. This issue was attributed to the lack of trained maintenance personnel. Special schools were set-up in country to address this issue. Pictured running at speed in Korea is a U.S. Army M46 tank. The M46 could attain a sustained maximum speed of 30mph (48.3kmh). *Patton Museum*

American M46 tanks were often brought to the crest of Korean hills as seen here to provide infantry fire support. The downside of this support activity was the mechanical abuse inflicted on the tanks by continuous climbing of steep hills that resulted in output shafts and final drives failing. *Patton Museum*

In a hull-down position is an American M46 tank during the Korean War. The large number of main gun rounds stored in the foreground indicates the tank is being employed in some type of stationary fire support role. The vehicle shown has the final version of the sheet metal covers over the engine exhaust mufflers. *Patton Museum*

Tankers are shown in Korea unloading 90mm main gun rounds from their shipping containers for their M46 tank parked behind them. The normal rate of consumption of 90mm main gun rounds for M46 tanks averaged 100 rounds per week per division during the Korean War. When used as artillery, that number could jump to 1,700 main gun rounds per week per division. *Patton Museum*

A U.S. Marine Corps M46 tank is pictured doing a bit of hill-climbing during the Korean War. The smaller of the two holes in the tank's gun shield is the opening for the .30 caliber (7.62mm) coaxial machine gun. The larger hole is for the gunner's magnified telescope sight. The gunner also had a magnified overhead periscope sight. *Patton Museum*

Rolling past the photographer is a U.S. Marine Corps M46 tank fitted with an unarmored 18-inch diameter Crouse-Hinds searchlight. Originally, the American military thought about fielding highly specialized searchlight tanks, however, it was more cost-effective to buy individual searchlights for each tank and risk losing them in battle then to fund the development and production of specialized searchlight tanks. *Patton Museum*

Shown is a U.S. Army M46 series tank fitted with a large infrared searchlight mounted on the right side of the gun shield. The infrared viewer is mounted on the front of the tank commander's cupola. The codename for the installation shown was Leaflet II. *Patton Museum*

Chapter Three:

M47 Patton Tank

The Communist North Korean invasion of its southern neighbor in June 1950 and the early success of its Soviet-supplied tank-led assault forces shook the senior levels of the American political and military leadership. They now became even more fearful of Soviet-sponsored military aggression around the world, but especially against Western Europe. Compounding the discomfort for the United States was the end of its atomic bomb monopoly. The Soviet Union-exploded its first atomic bomb on 29 August 1949. The continuing

Pictured is the first pilot model of the medium tank T42 armed with the T119 90mm main gun, which could withstand higher gun tube pressure and hence fire more powerful ammunition than the M3A1 90mm main gun on the M26 and M46 series tanks. The U.S. Army mounted the T42 turret on the M46 series tank hull to create the 90mm gun tank M47. *TACOM*

Prior to approving production of the M47 tank, a single pilot vehicle was created that mated a T42 tank turret onto an M46 tank chassis. The hybrid vehicle was designated the medium tank M46E1 and is shown here. *TACOM*

state of political and military tension between the Soviet Union and its client states and the United States and its various allies is referred to by historians as the Cold War (1946-1991).

Well aware that any Soviet military invasion of Western Europe would be spearheaded by its massive tank fleet, the U.S. Army realized that it required a great many modern tanks in the shortest amount of time possible to meet its new worldwide commitments. The conversion of the U.S. Army's inventory of M26 series tanks into M46 series tanks was merely a stopgap measure that would not provide the numbers of vehicles envisioned for its future needs. A

U.S. Army report completed in June 1953 called for an inventory of 18,617 tanks. Building brand new M46 series tanks was not desirable since many of its design features were rapidly becoming obsolete.

What was really needed was a new medium tank that reflected the latest design concepts. The problem was the long time frame implied in such an endeavor, which could take years. The U.S. Army's solution in July 1950 was to take the turret from an experimental vehicle designated the medium tank T42 (fitted with a new and advanced fire-control system) and mount on the chassis of the M46 series tank in order to come up with a suitable medium tank in the quickest amount of time. On 1 November 1950, this makeshift solution was designated as the 90mm gun tank M47 and ordered into full series production before a single prototype was built and tested. The U.S. Army term for this was a "crash-basis" production program. Also, their efficient use demanded gunners

In this picture we see a T42 turret on a redesigned M46 series tank hull with a new, more sloped front hull plate, referred to by tankers as the "glacis." The increased slope on the vehicle's glacis was achieved by eliminating the ventilating blower cover formerly located between the driver and bow gunner on the M46 series tanks. *TACOM*

A series production M47 tank is pictured during a training exercise. Notice the small, rounded protrusions on either side of the upper frontal portion of the turret. They are the vision ports for the tank's stereoscopic rangefinder operated by the vehicle's gunner. Sadly, the stereoscopic rangefinder took a great deal of training to master and was not popular with crews as a result. *Patton Museum*

with perfect (or at least identical) vision in both eyes, unlike co-incidence or other rangefinder types, as well as the fact that they did not work well in poor light conditions commonly found in western Europe.

Prior to the assembly of the first M47 tank, the U.S. Army took a single M46 series tank chassis and fit it with the experimental turret from the never fielded T42 medium tank and designated it the M46E1 tank. This vehicle was shipped to the U.S. Army's Aberdeen Proving Ground, Maryland, in March of 1951, were it was tested to identify any problems with it prior to series production. Actual

The most pronounced external feature that distinguished the M47 tank was the large extended rear turret bustle seen on the tank pictured. This design element also presented a serious problem with reentrant angles (shot traps) when the tank was engaged by enemy fire from the sides or rear. Notice the large stowage box attached to the rear of the tank's turret bustle. *TACOM*

series production of the M47 tank began in June 1951 before the testing process was completed. Aberdeen Proving Ground was the then home of the U.S. Army Ordnance Branch.

The use of the designation "gun tank" for the M47, rather than "medium tank," reflected a change in nomenclature by the U.S. Army in November 1950. It was concluded that the former weight bracket method of classifying American tanks as light, medium, or heavy was no longer viable due to changing concepts in the development and tactical employment of tanks. The M47 tank weighed in at 101,800lbs (46mt) combat loaded.

The design characteristics of the proposed M47 tank were outlined by the U.S. Army in January 1951 and received final approval in July 1951. A number of last minute design changes were incorporated into the new tank. The turret ring

An interior picture of a restored M47 tank belonging to the Military Vehicle Technology Foundation (MVTF) shows the vehicle's radio rack and radio with storage space for four main gun rounds above it. Behind the radio is the tank's electrically operated ventilation blower. *Chris Hughes*

diameter had to be increased from 69 inches (1.75m) found on the M46 series tank to 73 inches (1.85m) in order to accommodate the larger T42 turret. The suspension system on the M46 series tank had five track return rollers on either side of the hull, whereas the new tank would have only three track return rollers on either side of the hull.

Other changes to the planned M47 tank included the deletion of the ventilation blower as seen on the front hull roof of the M46 series tanks between the driver and assistant driver's positions. It was no longer needed because the turret of the new tank featured a ventilation blower in the roof of its turret bustle.

The ballistic protection level of the M47 tank was to be increased by sloping

This overhead picture from of an early production M47 tank illustrates some of the vehicle's external features, including the circular overhead portion of the vehicle's electrically operated ventilation blower, located at the rear of the turret bustle. *Patton Museum*

Taking part in a Western European training exercise is a platoon of U.S. Army M47 tanks. The vehicle rode on a torsion bar suspension system with 12 individually sprung dual road wheels (six per side) and six dual track return rollers (three per track). *Patton Museum*

the 4-inch (110mm) upper front hull (known as the glacis) at 60 degrees instead of the 46 degree slope found on the M46 series tanks.

The duplicate driver controls seen on the M46 series tank for the assistant driver/bow gunner was done away with on the M47 tank as the automatic transmission and power steering of the new tank made such a feature unnecessary.

As the first batch of series production M47 tanks rolled of the assembly lines, a number were sent to Aberdeen Proving Ground, Maryland and Fort Knox, Kentucky, home of the U.S. Army Armor Center, for an extensive testing process that lasted from August 1951 until August 1952. Despite the U.S. Army's hope that their new tank was combat ready, that proved not to be the case. The chassis of the M47 tank was reasonably reliable. However, the same could not be said for the vehicle's advanced fire-control system. It soon became clear from the testing process that at least fifteen key modifications to the M47 tank's fire-control system were required before it could even match the effectiveness of the much simpler fire-control systems found in the M26 and M46 series tanks.

The various modifications to the fire-control system of the M47 tank were quickly approved and applied to the vehicles already built and then introduced into the production lines. With the M47 tank finally being combat-ready, the bulk of the inventory went to U.S. Army armored units stationed in Western Europe. The U.S. Marine Corps would also employ the M47 tank. Production of the M47 tank would continue until November 1953 with a total run of 8,576 vehicles.

The U.S. Army would standardize the M47 tank on 22 May 1952. At the same time, the vehicle was officially nicknamed the "Patton II." However, at a later date the name was changed to the "Patton 47." Most American tankers simply referred to as the "forty-seven."

This picture shows the bow gunner's position in an M47 tank belonging to the Military Vehicle Technology Foundation. Unlike the bow gunner's position on the M46 series tank there was no duplicate driver's controls for the bow gunner on the M47 tank. *Chris Hughes*

In this photograph of the rear of an M47 tank, a number of features can be seen. At the bottom of the rear hull plate are three circular bolt-on covers, which allow access to the transmission. Just above the circular covers is the large square field telephone housing, while below the circular covers is a towing pintle. *Patton Museum*

A unit of U.S. Army M47 tanks is shown on a firing range in West Germany. Clearly visible on the vehicles shown are the sheet metal covers over the engine's right hand side exhaust mufflers. They were installed to minimize the risk of burns to the crew as the mufflers tended to become extremely hot when the vehicle was operated for extended periods. *Patton Museum*

The crew of a U.S. Army M47 tank is shown carefully backing up their vehicle into a U.S. Navy landing craft. The M47 tank was 27 feet 11 inches (8.5m) long and had a width of 11 feet 6.3 inches (3.5m). The vehicle was 10 feet 11 inches (3.3m) tall and had a combat weight of 101,800 pounds (46,175.5kg). *Patton Museum*

This M47 tank is pictured on display at the former Patton Museum of Armor and Cavalry. Notice the cylindrical blast defector. The tank carried 71 main gun rounds for its M36 90mm cannon and 11,150 rounds of ammunition for its two .30 caliber (7.62mm) machine guns. *Richard Hunnicutt*

Looking down through the loader's hatch of an M47 tank the breech of the 90mm main gun is visible in the picture, as well as the recoil guards on either side of the breech. To the left of the main gun is the coaxial .30 caliber (7.62mm) machine gun. The turret of the M47 tank could be turned 360 degree in 15 seconds using its electric-hydraulic traversing system. *Chris Hughes*

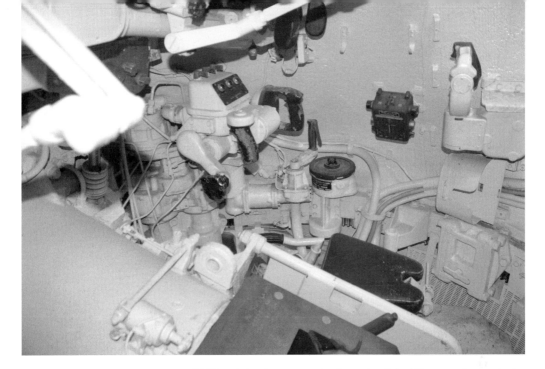

From the loader's position in an M47 tank looking over the breech of the 90mm main gun is the vehicle's gunner's position. Just in front and above the gunner's seat is the circular azimuth indicator, which is used to set off azimuth angles when laying the main gun for indirect fire, or for planned night direct fire targets. *Chris Hughes*

A picture taken from the loader's position on an M47 tank shows the vehicle commander's seat, which is located behind and just above the gunner's seat that is visible in the lower left portion of the picture. The tank commander stood on his seat when looking out over the top of his cupola. *Chris Hughes*

This U.S. Army M47 tank is on field maneuvers. There were 1,700 rounds of ammunition carried onboard the vehicle for the .50 caliber (12.7mm) machine gun that was mounted on the turret roof. The weapon had a rate of fire of 450 to 550 rounds per minute. *Patton Museum*

The driver of an M47 tank is shown wearing the Second World War American-designed and manufactured leather crash helmet and dust goggles. Notice the tank commander's pintle-mounted .50 caliber (12.7mm) machine gun has been replaced with a .30 caliber (7.62mm) machine gun. *Patton Museum*

This picture shows a view of the driver's position on a late production M47 tank. Notice the numerous fire extinguishers stored in the front hull of the vehicle. On a level road the driver could sustain a speed of 30 miles per hour (48.3kmh). *Chris Hughes*

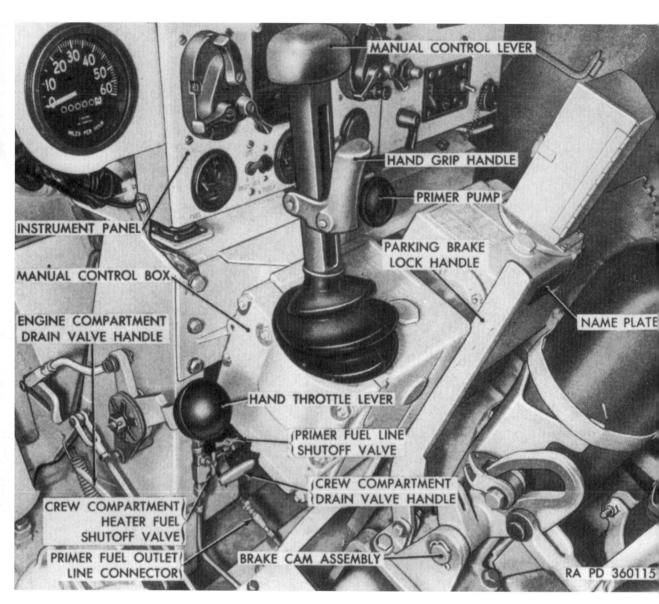

MANUAL CONTROL LEVER

HAND GRIP HANDLE

PRIMER PUMP

PARKING BRAKE
LOCK HANDLE

NAME PLATE

INSTRUMENT PANEL

MANUAL CONTROL BOX

ENGINE COMPARTMENT
DRAIN VALVE HANDLE

HAND THROTTLE LEVER

PRIMER FUEL LINE
SHUTOFF VALVE

CREW COMPARTMENT
DRAIN VALVE HANDLE

CREW COMPARTMENT
HEATER FUEL
SHUTOFF VALVE

PRIMER FUEL OUTLET
LINE CONNECTOR

BRAKE CAM ASSEMBLY

RA PD 360115

An illustration from an M47 tank manual shows the driver's controls. He steered the tank using the single lever shown referred to as a "wobble stick." The wobble stick operated a hydraulically actuated controlled differential. *Patton Museum*

A group of U.S. Army soldiers are shown removing the power pack of an M47 tank, which consisted of a Continental 12-cylinder air-cooled, gasoline-powered, AV-1790-5B engine and an Allison Cross-Drive CD-850-4 transmission. The vehicle's engine is on the left and the transmission is on the right. *Patton Museum*

The five-man crew of an U.S. Army M47 tank displays for the photographer the many contents of their tank. Among the many items visible in the picture are the tank's three machine guns as well as the crew's flashlights and headphone sets. *Patton Museum*

A U.S. Army M47 tank is shown conducting a fording test in the Fulda River near Fulda, Germany on December 11, 1953. All postwar U.S. Army tanks were required to ford, without any prior preparation, a 48 inch deep (1.22m) water obstacle. *Patton Museum*

The crew of a U.S. Army M47 tank is trying to cross a field obstacle. The vehicle could handle a grade of 60 percent and cross over a trench 8.5 feet (2.6m) wide. Like all American postwar tanks, it had the ability to climb a vertical wall 36 inches (91.4cm) tall. *Patton Museum*

From a U.S. Army manual comes this picture of an 18-inch diameter Crouse-Hinds searchlight mounted on the gun shield of an M47 tank. This was the same searchlight mounted on the M46 series tank during the Korean War. *Patton Museum*

Pictured is a U.S. Army M47 tank that is taking part in a field training exercise near Mount Fuji, Japan in February 1956. Unlike the suspension system on the M46 series tanks that had five track return rollers, the M47 tank only had three. *Patton Museum*

This picture reflects an early U.S. Army experiment with mounting smoke grenade launchers on their armored fighting vehicles, in this case an M47 tank. This tank has smoke grenade launchers on the front, sides, and rear of the turret oriented to fire horizontally rather than the front turret mounted smoke grenades launchers, which are typically angled to fire upwards, as has been common for many decades.. *TACOM*

The fate of many obsolete tanks is to be consigned to target ranges for weapon tests. Pictured is an M47 tank just struck by some sort of large antitank missile. The M47 tank in the foreground has a T-shaped blast defector, which appeared very late in the production cycle. *DOD*

Belonging to a private American collector is this M47 tank that is missing its fenders. Notice the fender support brackets on the right side of the vehicle's hull. The rain gutters and the various handholds on the tank's turret mark it as a former West German Army vehicle. *Michael Green*

An overhead picture of an M47 tank upgraded with an M68 105mm main gun and a diesel engine by an Italian firm interested in drumming up business with countries that might wish to modernize their tank inventories. *Patton Museum*

Chapter Four:

M48 through M48A3 Patton Tanks

The U.S. Army expressed its interest to the Chrysler Corporation in the development of a successor vehicle to the M47 tank in late 1950. In order to speed up the developmental process of the new tank, it was to be powered by the same engine and transmission found in the M47 tank. However, rather than retain the external hull design of the earlier tank, the U.S. Army decided to use the elliptically shaped hull of the 1950 design for the Heavy Tank T43, which in April 1956 was designated the 120mm gun combat tank M103.

An early model 90mm gun tank T48 is pictured undergoing mobility testing at Fort Knox, Kentucky. Early model T48 tanks can be identified by their cylindrical blast deflector and the lack of track tension idlers located between the last road wheels and the rear-mounted drive sprockets. *Patton Museum*

The second pilot of the 120mm gun tank T43E2 is shown. After further modifications it was designated the 120mm gun, full tracked, combat tank M103A1. The elliptically shaped cast homogenous armor (CHA) hull for the T48 tank was copied from the M103A1 tank as was the upper rear hull design. *Patton Museum*

Employing cast homogenous armor (CHA) elliptically shaped hull and a new hemispherical CHA turret on the next generation medium tank permitted the maximum ballistic armor protection level for a given volume of space with a minimum of weight.

With the adoption of the elliptically shaped hull from the T43 tank came a decrease in crew size on the proposed new medium tank from five to four. Done away with was the assistant driver/bow gunner's position in the vehicle's front hull. This position dated back to the U.S. Army's M2 combat cars/light tanks of the 1930s. Combat experience from the Second World War revealed that the assistant driver/bow gunner had little combat effectiveness and combined with a desire to increase armor protection levels led to the adoption of the ballistically superior elliptically shaped front hull and the elimination of the position on the T43 tank and the replacement vehicle for the M47 tank.

The U.S. Army contracted with the Ordnance Development Department of the Chrysler Corporation to take the general outline of what they wanted in a new medium tank and come up with an acceptable vehicle that could then be placed into series production as soon as possible. Chrysler's first task was to complete the detailed design of the new tank and build six pilot vehicles, five for testing by the U.S. Army and one for testing by the U.S. Marine Corps. A pilot vehicle is intended to prove that the manufacturing line works and is able to actually produce the vehicle in numbers.

Chrysler quickly came up with ½ scale clay design model of the hull and turret of what they believed the U.S. Army wanted in a new tank. The clay model met the approval of the U.S. Army on 2 February 1951. Advance drawings based on the clay model soon went out to foundries for them to make the necessary armor castings.

It took the U.S. Army until 27 February 1951, before the proposed new medium tank project was officially initiated and its characteristics outlined. The new vehicle was designated the 90mm gun tank T48 and armed with a new light-weight 90mm main gun originally designated the T139, which matched the ballistic performance of the heavier M36 90mm main gun mounted in the M47 tank. The T139 gun tube also had a quick change feature that expedited its removal in the field when the need arose. The T139 main gun was later standardized as the M41.

The large CHA hemispherical shaped turret of the T48 tank is clearly evident in this picture showing it parked next to an M47 tank. The rounded shape of the T48 tank turret was made possible by the adoption of the 85 inch (2.16m) diameter turret ring of the T43 tank. The turret ring diameter of the M47 tank was only 73 inches (1.85m). *Patton Museum*

The turret walls of the T48 tank sloped smoothly down to the top of the vehicle's hull, eliminating most reentrant angles (shot traps). T48 tanks did not have a rear turret bustle rack. Instead, there were two horizontal bars to which items could be attached to. Notice the engine exhaust muffler just under the rear turret overhang. *Patton Museum*

The T48 tank boasted two coaxial machine guns; a .50 caliber (12.7mm) on the left of the main gun and a .30 caliber (7.62mm) on the right of the main gun. Unfavorable test results soon led to the removal of the .50 caliber (12.7mm) machine gun on the left side of the main gun and its replacement with a more suitable .30 caliber (7.62mm) machine gun. The .30 caliber (7.62mm) machine gun on the right side of the main gun was replaced with a direct sight telescope for the tank's gunner.

The T48 tank also featured a Chrysler-designed remote control machine gun mount attached to the tank commander's cupola, which allowed him to fire his .50 caliber (12.7mm) machine gun from within the confines of the vehicle's turret. However, reloading the weapon required the tank commander to expose his head and upper torso out of his cupola.

Testing of the T48 tanks began in February 1952 and continued until the end

During testing of the T48 tank it was discovered that the heat from the rear hull roof mounted exhaust muffler would freeze the gun travel lock in place when the main gun was locked in the travel position as seen in this picture. The gun travel lock would become so hot that it could only be touched with asbestos gloves. *Patton Museum*

of 1952. Sadly, the constant perceived threat of Soviet aggression in Western Europe impelled the U.S. Army senior leadership to rush the T48 tank into series production before the inevitable bugs could be worked out of the new tank. Instead, it was decided that any needed design changes uncovered by the testing of the T48 tanks would be incorporated into the series production vehicles as quickly as possible. The T48 tank weighed in at 98,400lbs (45mt) combat loaded.

The first series production T48 tanks came off the Chrysler assembly line in April 1952. To meet the demand for large numbers of T48 tanks, both the Ford Motor Company and the Fisher Body Division of General Motors Corporation were brought into the program. Even as numbers of T48 tanks rolled off the production lines, continuous testing of the vehicles uncovered an ever increasing list of major and minor design problems that had to be addressed. Things became so problematic with the T48 tank that the U.S. Army decided in January

1953, that the vehicle was unfit to be shipped overseas for use by our frontline units. Instead, the T48 tank was suitable only for in-country training duties, if a number of major design shortcomings were addressed first.

Early production T48 tanks lacked the track tension idlers seen on the M46 and M47 series tanks, which were located between the rear set of road wheels and the rear hull-mounted drive sprockets. However, a decision was made early-on to reinstall them on series production vehicles.

Testing had uncovered the fact that the original driver's hatch design of the T48 tank proved to be too small and made for an uncomfortable seating position for the driver when operating the tank with his hatch in the open position.

To solve the problem of the overheated gun travel lock on the T48 tank and follow on vehicles, a pair of curved metal deflectors were installed just behind the engine exhaust muffler to direct the hot exhaust gases away from the travel lock. Those deflectors can be seen in this picture. *Michael Green*

An early model T48 tank climbs over a 36 inch (91.4cm) step at Aberdeen Proving Ground, Maryland, former home of the U.S. Army Ordnance Corps. Testing of the T48 tank revealed that its turret and fire control system did not function properly. The U.S. Army decided to await future turret and fire control developments and continued with production of the vehicle. *Patton Museum*

To rectify this problem a new larger driver hatch was devised that was introduced into series production vehicle as early as possible.

By early 1953, production of the T48 tank was in full swing with almost 900 vehicles completed by March. On 2 April 1953, the U.S. Amy standardized the T48 tank as the 90mm gun tank M48. At the same time, the new tank was officially nicknamed the "Patton 48" in honor of General George S. Patton, however, most American tankers simply referred to them as the "forty-eights" or "M48s." Roughly 3,200 units of the M48 tank rolled off the assembly line.

A production run of 120 early-model M48 tanks were built with defective hull armor and were designated the M48C tank and were to identified by markings on the front hull glacis as "non-ballistic training tank only."

On 25 October 1954, the U.S. Army decided to amend its designation of the M48 tank. Thereafter, the tanks with the original small driver's hatch and the Chrysler-designed remote control machine gun mount attached to the tank commander's cupola would remain the 90mm gun tank M48. However, all those tanks with the larger driver hatch and fitted with a new small turreted cupola armed with a .50 caliber (12.7mm) machine gun and designed by Aircraft

A T48 tank races down a steep hill. Mobility testing of the T48 tank and the constant throwing of tracks led the U.S. Army to decide to add a track tension idler to the later production versions of the vehicle as was seen on the M46, M46A1 and M47 tanks. The T48 rode on a torsion bar suspension system. *Patton Museum*

Early models of the T48 tank had a very small driver's hatch design as seen from this picture. The driver's periscopes automatically retracted before the hatch swung open. However, when the driver closed his hatch he had to manually push the periscopes back into position. For this reason and others, a new larger driver's hatch was devised. *Patton Museum*

Armaments Incorporated would be designated as the 90mm gun tank M48A1. The new tank commander cupola would allow the tank commander to fire and reload his weapon from within the confines of his tank's turret. Originally referred to as the .50 caliber (12.7mm) machine gun mount M1 it was eventually referred to as the M1 cupola. Besides entering into U.S. Army service the M48A1 tank took the place of the M47 tank in U.S. Marine Corps service.

Because of the smaller turret roof opening for the Chrysler-designed remote control machine gun mount those M1 cupolas retro-fitted to early production M48A1 tanks came with an adaptor ring.

The U.S. Army never wavered in its efforts to perfect the M48 series tanks for

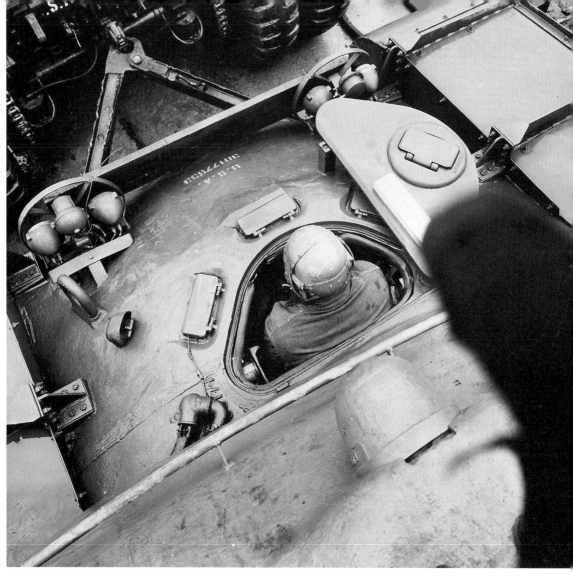

A new, larger, driver's hatch was soon developed for the T48 tank as seen in this photograph. The new hatch raised up and then swung open to the right and rested on a very small metal pillar. Unlike the original small driver's hatch the periscopes did not retract and remained fixed in position on the larger hatch design. *Patton Museum*

service use. The 104,000 pound (47mt) combat loaded M48A1 tank was soon superseded by another improved version approved for series production in December 1955 and designated the 90mm gun tank M48A2. A total of 2,328 M48A2 tanks rolled off the assembly line.

A distinguishing external feature of the M48A2 tank was a raised rear engine deck combined with a vertical rear hull plate featuring two large louvered doors that replaced the complicated grill work seen on the rear engine decks of the M48 and M48A1 tanks. This new rear hull arrangement on the M48A2 tank helped in cooling the engine and minimizing the infrared (heat) signature of the

The T48 tank became the 90mm gun tank M48 in April 1953. Pictured at Fort Hood, Texas is an M48 tank. The original cylindrical blast defector was eventually replaced with the T-shaped blast defector seen here. This vehicle also has the Chrysler-designed low profile tank commander cupola seen on all the T48 and M48 tanks. *Michael Green*

vehicle and was carried on through all the follow-on versions of the M48 series tanks.

In place of the rounded fender extensions seen on the front and rear of the M48 and M48A1 tanks, the M48A2 tank had squared fender extensions, with the front fenders extension having a horizontal cross bar added for extra strength. The M48A2 tank also featured new squared front hull-mounted brush guards around the identical headlight clusters. This was in contrast to the rounded brush guards around the front hull mounted non-identical headlight clusters seen on the M48 and M48A1 tanks.

U.S. Army unhappiness with the Chrysler-designed tank commander cupola on the T48 and M48 tanks led to the adoption of a new enclosed tank commander cupola seen here on a T48 tank armed with an internally-mounted .50 caliber (12.7mm) machine gun and eventually designated the M1 cupola. Tanks equipped with the new cupola were designated the M48A1 tank. *Patton Museum*

The M48A2 tanks featured only three track return rollers on either side of the tank's hull. The M48A2 tank weighed in at 105,000lbs (48mt) combat loaded.

A key internal improvement to the M48A2 tank was the use of a new fuel injected gasoline engine referred to as the Continental AVI-1790-8. The new engine offered increased horsepower and better fuel economy. As an added perk, the new engine was smaller than its predecessors allowing for an increase in internal fuel capacity, which extended the vehicle's cruising range on roads to 160 miles (257km). By way of contrast, the M48 and M48A1 tanks had an approximate road range of only 70 miles (113km). Prior to fuel injected engines tank engines were fed fuel by carburetors.

Testing of the M48A2 tank and field use soon uncovered some unforeseen problems with the vehicle, which were corrected on 1,344 units of the vehicle already produced. These modified tanks were assigned the designation M48A2C. A key external identifying feature of the M48A2C tank was the deletion of the track tension idlers. An important internal change to the M48A2C tank was the replacement of the stereoscopic rangefinder as found on the M47 through M48A2 tanks and the mounting of a coincidence rangefinder in its place. The U.S. Marine Corps never took the M48A2 or M48A2C tanks into service.

By the time production of the M48 series tanks ended in 1959, almost 12,000

An overhead view of a T48 tank with the M1 cupola fitted. In the rear left corner of the turret is the circular overhead portion of the electrically-powered ventilation blower that was provided for crew comfort. The T48 tank had storage space for 60 main gun rounds. *Patton Museum*

A somewhat worse for wear M48A1 tank. Notice the adaptor ring below the cupola that marks it as a tank that was originally fitted with the Chrysler-designed cupola. Like the T48 tank, the M48A1 tank had storage space for 60 main gun rounds. It also carried 500 rounds of .50 caliber (12.7mm) ammunition and 5,900 rounds of .30 caliber (7.62mm) ammunition. *Michael Green*

units had been built. Much to the dismay of the U.S. Army, the short operational range of the early M48 series tanks and their numerous design problems had made them a major object of American congressional investigations in the late 1950s.

The U.S. Army's long aversion to employing diesel engines in its ground vehicles changed in June 1958, when the use of diesel fuel was to be allowed, if it significantly contributed to better fuel economy. This decision and the need for a new more powerfully armed tank to counter the introduction of the low-slung and thickly-armored Soviet Army T54 medium tank pushed the U.S. Army to develop a new tank as a replacement for the M48 series tanks.

In 1959, the U.S. Army modified an M48A2 tank with a new front hull configuration, a Continental AVDS-1790-2 air-cooled diesel engine, and a British-designed but American-modified and built 105mm main gun designated the M68. To highlight the improvement over the earlier gasoline-engine powered M48 series tanks, on 16 March 1959 the new vehicle was designated as the 105mm gun full tracked combat tank M60.

By the time production of the M60 series tanks tank ended in 1982, over 15,000 units had been built in four different versions, including the original M60, the M60A1, the M60A2, and the M60A3. M60 tank production was so high at this time because the USA chose to sell to foreign buyers for only £ 65,000 each (i.e. £ 6,000 below cost price) in order to maintain the production facilities and specialised work force. The final version of the M60 series tank weighed in at 114,600lbs (52mt) combat loaded. Although the M60 tank series was obviously just a product improved version of the earlier M48 tank series, they were never officially or un-officially referred to as "Patton tanks."

Even before production of the M60 series tanks began, the U.S. Army was exploring the possibility of upgrading its inventory of M48A1 tanks with a Continental AVDS-1790-2 air-cooled diesel engine and a M68 105mm main gun. A single pilot vehicle was modified with the air-cooled diesel engine and a M68 105mm gun and designated the M48A1E1. However, the inability to mount either daylight or infrared vision devices in the already cramped M1 cupola on

This picture taken during a field exercise shows an M48A1 tank in the foreground and an M48 tank in the background. The M48A1 tank was provided with an opened-topped enclosed turret bustle rack, which normally contained items like the crew's sleeping bags and spare uniforms.
Patton Museum

This photograph of a U.S. Marine Corps M48A1 tank with its turret reversed, shows the main gun mounted in the gun travel lock, which supports the gun (near the blast defector) when the vehicle is traveling and not in a combat situation. There was no stabilizer system on the M48A1 tank's main gun. *RC McInteer collection*

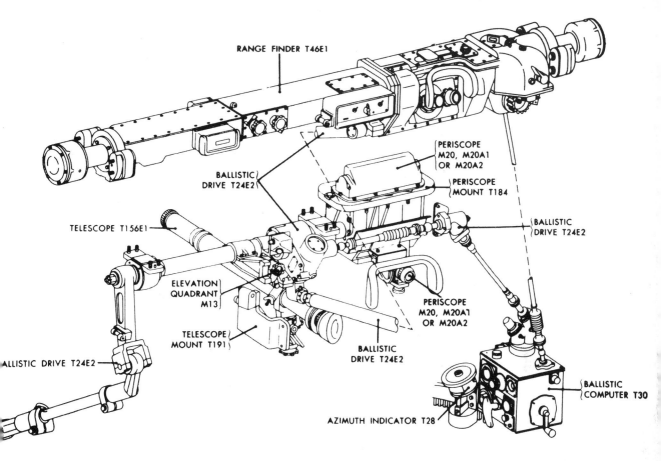

RANGE FINDER T46E1

BALLISTIC DRIVE T24E2

PERISCOPE M20, M20A1 OR M20A2

PERISCOPE MOUNT T184

BALLISTIC DRIVE T24E2

TELESCOPE T156E1

ELEVATION QUADRANT M13

PERISCOPE M20, M20A1 OR M20A2

TELESCOPE MOUNT T191

BALLISTIC DRIVE T24E2

BALLISTIC DRIVE T24E2

BALLISTIC COMPUTER T30

AZIMUTH INDICATOR T28

This line illustration from a U.S. Army manual shows the arrangement of components that made up the Phase IV fire control system on the M48 series tank, which included the stereoscopic rangefinder T46E1 and the ballistic computer T30. Unlike the M47 tank where the gunner operated the rangefinder, on the M48 tank it was operated by the tank commander. *Patton Museum*

the M48A1 tank put a stop to that effort. The effort was rekindled in 1961 when approval was given to upgrade 600 M48A1 tanks with new the AVDS-1790-2 diesel engine.

Instead of getting an M68 105mm main gun, the upgraded M48A1 tanks were to retain their existing M41 90mm main guns for two reasons. First, funding shortfalls made it impossible to supply enough 105mm main gun rounds for all the planned converted tanks. Second, there were still large stockpiles of 90mm main gun rounds on hand. The newest version of the M48 series tank received the designation 90mm gun full tracked combat tank M48A3 and weighed in at 107,000lbs (48mt) combat loaded.

The first M48A3 tank was accepted by the U.S. Army in February 1963. The biggest improvement that this latest version of the M48 series tank series brought to the table was its greatly improved operational range on roads that went up to 300 miles (483km). This came about due to larger fuel tanks and the greater thermal efficiency of diesel fuel.

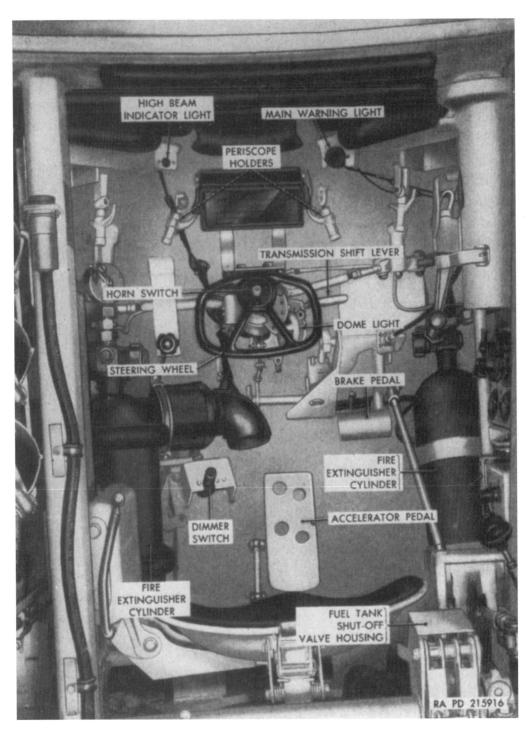

HIGH BEAM INDICATOR LIGHT

MAIN WARNING LIGHT

PERISCOPE HOLDERS

TRANSMISSION SHIFT LEVER

HORN SWITCH

DOME LIGHT

STEERING WHEEL

BRAKE PEDAL

FIRE EXTINGUISHER CYLINDER

DIMMER SWITCH

ACCELERATOR PEDAL

FIRE EXTINGUISHER CYLINDER

FUEL TANK SHUT-OFF VALVE HOUSING

RA PD 215916

From an M48A1 tank manual appears this picture of the driver's compartment with all the various components and features labeled. With automatic transmissions and power steering, the M48 tank series was much easier to drive than its foreign counterparts such as the British Centurion series tanks and the Russian T54/55 series tanks. *Patton Museum*

The original stereoscopic rangefinder mounted in the M48A1 tank was replaced on the M48A3 tank with a coincidence rangefinder. The M48A3 tank also received the Nuclear/Biological/ Chemical (NBC) gas particulate unit from the M60 series tanks.

Besides 600 M48A3 tanks for the U.S. Army, another 419 M48A1 tanks were converted to the M48A3 standard for the U.S. Marine Corps between 1963 and 1965. A key external spotting feature of the M48A3 tank were the five return rollers on either side of the tank's hull in lieu of the three seen on the U.S. Army's M48A2 and M48A2C tanks.

Another external spotting feature of the M48A3 tanks (not seen on any of its predecessors) were the large dry type (paper) air cleaner boxes mounted on either side of the vehicle's hull over the rear portion of the vehicle's horizontal

Pictured is an overturned U.S. Marine Corps M48A1 tank. Driving any large tracked vehicle on side slopes can be extremely difficult as they have a tendency to drift downhill. It is up to the driver to perform constant corrections to keep the vehicle straight. *RC McInteer collection*

A shortcoming of the M48 and M48A1 tanks was their limited operational range of about 70 miles. This was a consequence of small internal fuel tanks and their high consumption gasoline engines. The U.S. Army's partial answer to the problem was the adoption of external fuel tanks seen perched on the rear hulls of the M48A1 tanks pictured. *Patton Museum*

fenders. These were the same dry type air cleaner boxes seen on the horizontal fenders of the M60 series tanks. Prior to the addition of the external dry type air cleaners on the M48A3 tanks, all of its predecessor M48 series tanks had interior oil bath air cleaners.

An easy to miss external spotting feature of the M48A3 tank was the crew compartment heater exhaust pipe that passed through the hull roof on the right side of the driver's hatch and extended out to the right side of the tank. This feature is also seen on the M60 series tanks. On the predecessor M48A2 and M48A2C tanks the crew compartment heater exhaust pipe passed through the hull roof on the left side of the driver's hatch and extended out to the left side of the tank. On the M48 and M48A1 tank there were two parallel crew compartment heater exhaust pipes coming out on the left side of the driver's hatch and extending out to the left side of the tank.

This picture shows the arrangement of the four unarmored 55 gallon fuel drums that were attached to rear of the M48A1 tank and connected to the vehicle's fuel system. The additional fuel supply nearly doubled the operational range of the M48A1 tank to 135 miles. *Patton Museum*

In 1967, another 578 M48A1 tanks were converted into M48A3 tanks. This second batch of upgraded tanks sported some new features not seen on the original M48A3 tanks and were therefore referred to as M48A3 (Mod B) configuration.

A key external spotting feature of the M48A3 (Mod B) tanks was a spacer ring (designated as the G305 turret cupola riser) that contained nine large vision blocks and was installed between the turret roof and the M1 cupola in order to improve the tank commander's visibility. The M1 cupola on the Mod B configuration also came with a new bulged hatch cover to provide the tank commander with more headroom.

U.S. Marine Corps M48A3 tanks had received the G305 turret cupola riser as a field modification before the U.S. Army M48A3 tanks went through the Mod B upgrades.

Also seen for the first time with the Mod B configuration of the M48A3 tank was a heavy metal gauge turret interrupting bar mounted in front of the M1 cupola. Its intended role was to prevent the vehicle commander from firing into the rear of the Xenon 2.2 kilowatt white light/infrared searchlight that was often mounted over the top of the gun-shield on the Mod B configured M48A3 tanks.

Late production M48A3 (Mod B) tanks came off the production lines with an embossed X-shaped pattern on their squared front and rear fenders extensions. This was in lieu of the horizontal cross bar added for extra strength on the squared front fenders extensions of the M48A2 tank through the early production (M48A3) Mod B tanks. These embossed X-shaped patterns also appeared on the frontal and rear portions of the horizontal fenders.

Based on user feedback late-production M48A3 (Mod B) tanks came with new thicker and stronger square brush guards around the vehicle's two rear tail lights. Another external improvement on late production M48A3 (Mod B) tanks was the welded on of a thick protruding horizontal steel bar around the top of the two large louvered doors at the rear of the vehicle hull. When early production M48A3 tanks eventually returned for rebuilding they received all the Mod B improvements and the Mod B designation was then dropped.

Shown without the four unarmored 55 gallon fuel drums fitted is the metal rack that attached to the rear of the M48A1 tank. If the need arose, the metal rack could be jettisoned from within the vehicle. *Patton Museum*

A—CARBURETOR ADJUSTMENT OPENING
B—MANUAL STARTING HANDLE
C—LIFTING EYE
D—EXHAUST OUTLET
E—AUXILIARY GENERATOR AND ENGINE NAME PLATE
F—NEGATIVE TERMINAL
G—FUEL LINE QUICK DISCONNECT
H—COOLING AIR INLET
J—POSITIVE TERMINAL
K—GUIDE RAILS
L—MAGNETO ADJUSTMENT HOLE COVER
M—TIMING HOLE COVER
N—CARBURETOR AIR INLET
P—CONTROL CABLE RECEPTACLE COVER

RA PD 275942

Taken from a U.S. Army manual is this picture of the combined auxiliary generator and engine located in the engine compartment of all gasoline-engine powered M48 series tanks. Nicknamed the "Little Joe," the 1-cylinder, 4-cycle, gasoline engine produced 14.5 horsepower and powered the vehicle's electrical system when the main engine was turned off. *Patton Museum*

Somebody felt that doing away with the bow gunner on the M48 series tanks was a mistake. To rectify this perceived problem, machine gun pods (armed with .50 caliber (12.7) machine guns) were developed and mounted on the front fenders of the M48A1 tank pictured. Nothing ever came of this project. *Patton Museum*

A distinguishing feature of the M48 and M48A1 tanks were their rounded front and rear fender extensions and the two slightly different headlight clusters mounted on the front hulls of the vehicles. These headlight clusters were protected by large rounded brush guards as seen in this picture. *Michael Green*

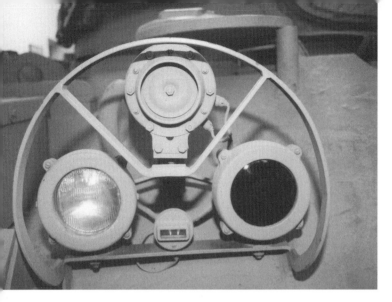

The headlight cluster pictured has both the white light headlight and the infrared headlight. Above the headlights is a horn and below the horn is blackout marker light. *Michael Green*

Displeasure with the very limited operational range of the M48 and M48A1 tanks pushed the U.S. Army to field the 90mm gun tank M48A2 pictured here. It came with a new, more compact Continental 12 cylinder air-cooled fuel-injected gasoline engine designated the AVI-1790-8, which allowed for larger internal fuel tanks and pushed the tank's operational range to about 160 miles. *TACOM*

An external spotting feature on the front of the M48A2 tank was new squared front fender extensions with a cross-bar bracket to strengthen them. Also visible in this image are the new identical front hull-mounted headlight clusters that have a more squared brush guard. To save weight, M48A2 tanks had only three track return rollers. *TACOM*

A close-up photograph of one of the two identical headlight clusters that first appeared on the front hulls of M48A2 tanks. This same headlight cluster arrangement and brush guard would remain on all the follow-on vehicles of the M48 tank series. *Michael Green*

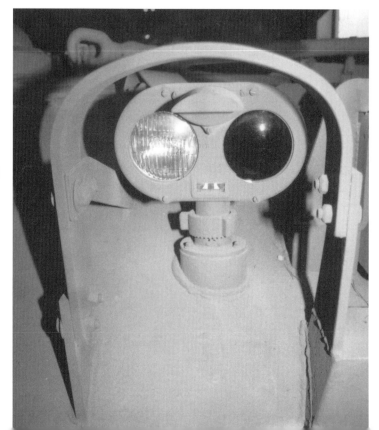

A big problem with the M48 and M48A1 tanks was their large infrared signature. The solution shown on the M48A2 tank pictured was a redesigned and enlarged rear engine compartment that significantly cut down on the vehicle's infrared signature. Also visible are the horizontal air intake grills along each side of the solid center section. *Patton Museum*

This picture shows the redesigned and enlarged rear engine compartment of an M48A2 tank. The solid center section of the rear hull roof compartment covered an insulated exhaust tunnel that discharged the engine's exhaust gases out the rear of the tank's hull. Notice the semi-enclosed turret bustle rack that first appeared on the M48A1 tank. *Patton Museum*

This photograph shows some of the main gun rounds fired from the M41 90mm main gun on the M48 series tanks. On the left are two kinetic energy rounds. In the middle is a high explosive plastic (HEP) round. To the right of the HEP round is an anti-personal canister round. On the far right is a high-explosive (HE) round with a fuze that can be set to detonate on contact or as a result of a delay action. *Michael Green*

Pictured is an M48A2 tank fitted with a mount for five French-designed SS-11 wire-guided anti-tank missiles, although only three are in this photograph. These missiles weighed about 60 pounds (27kg) each and had an effective range up to 3,280 yards (3,000m). This experimental arrangement proved impractical and was not pursued. *TACOM*

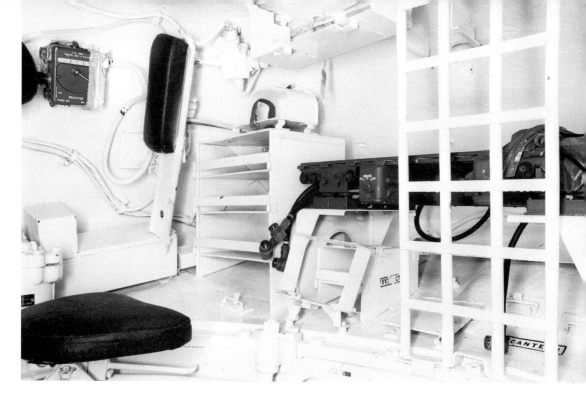

From the loader's position in an M48A2 tank can be seen the radio rack (minus its radio) located in the rear of the vehicle's turret. Next to the radio rack is a shelf arrangement intended to store spare periscopes. Also visible is the tank commander's seat and backrest. The metal framework in front of the radio rack is to deflect main gun cartridge cases as they are expelled from the breech of the main gun in the counter-recoil movement. *Patton Museum*

The U.S. Army continued to improve the M48A2 tank with some fire-control upgrades. The slightly improved tank was designated the 90mm gun tank M48A2C. The only external difference between the M48A2 and the M48A2C tank was the deletion of the track tension idlers seen on this M48A2C tank belonging to the former Patton Museum of Armor and Cavalry. *Michael Green*

Pictured on the rear hull of this M48A2C tank that has managed to get itself stuck on a frozen rice paddy dike in South Korea are the two large louvered doors that deflect the engine exhaust gases horizontally, thereby improving the vehicles infrared signature. *Patton Museum*

Even before the first M60 tank rolled off the production line in 1959, the U.S. Army began thinking about modernizing its inventory of obsolete M48A1 tanks with M60 tank components. One of six pilot vehicles used to test the feasibility of the concept is pictured here and was designated the 105mm gun full tracked combat tank M48A1E1. *TACOM*

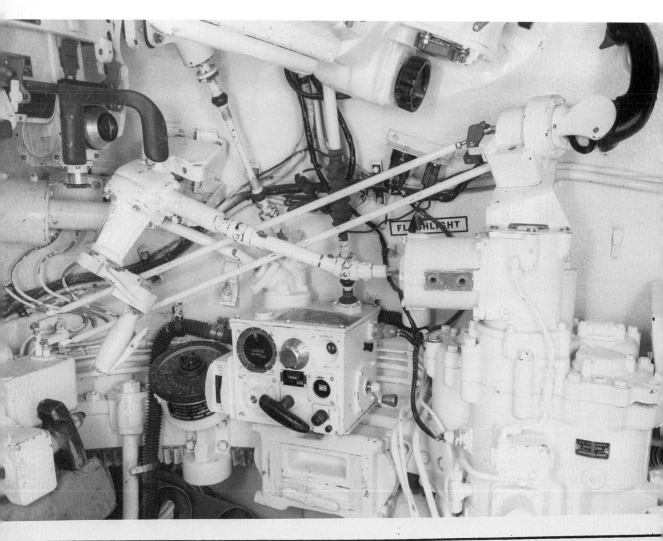

FIGURE 24
RIGHT SIDE OF TURRET SHOWING NEW COMMANDER'S HANDLE, BRACKET, AND RODS; RELOCATED
MANUAL TRAVERSE AND SUPERELEVATION ACTUATOR; AND MODIFIED COMPUTER M13A3 — M48A2C TANK

LR NU34-3

Prepared for:	ARMY MATERIEL COMMAND	3-17-66	66-405
By:	CHRYSLER CORPORATION DEFENSE ENGINEERING	Date	Negative

The gunner's position on the M48A2C tank is shown with the new M13A1C ballistic computer located in the center of the picture. The stereoscopic rangefinder found on the M47 tank through the M48A2 tank was replaced with a coincidence rangefinder on the M48A2C tank, which proved much easier to train people on than the stereoscopic rangefinder. *Patton Museum*

Tests conducted with the M48A1E2 tank were successful and the vehicle was re-designated as the 90mm gun full tracked combat tank M48A3. Rather than the three track return rollers of the M48A1E2 tank, the series production M48A3 tanks had five track return rollers as seen on the vehicle pictured. *TACOM*

Eventually the U.S. Army decided to stick with the original M41 90mm main gun on the M48A3 tank. Here we see a picture of one of the two M48A1E2 tanks that was re-gunned with the 90mm main gun and then re-designated as the 90mm gun full tracked combat tank M48A1E2. This tank has the adaptor ring under the M1 cupola that identifies it as once having been fitted with the original Chrysler-designed cupola. *TACOM*

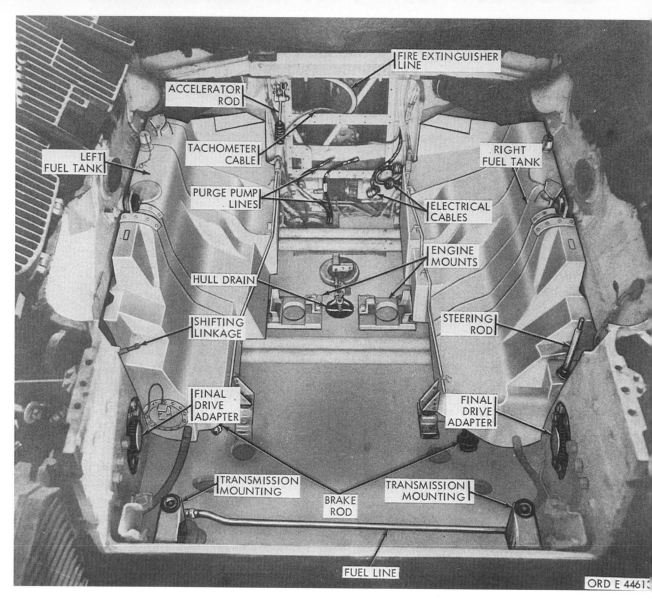

FIRE EXTINGUISHER LINE

ACCELERATOR ROD

LEFT FUEL TANK

TACHOMETER CABLE

RIGHT FUEL TANK

PURGE PUMP LINES

ELECTRICAL CABLES

ENGINE MOUNTS

HULL DRAIN

SHIFTING LINKAGE

STEERING ROD

FINAL DRIVE ADAPTER

FINAL DRIVE ADAPTER

TRANSMISSION MOUNTING

TRANSMISSION MOUNTING

BRAKE ROD

FUEL LINE

ORD E 4461

The process of modernizing M48A1 tanks resulted in the first M48A3 tank being accepted by the U.S. Army in February 1963. Taken from a U.S. Army manual is this picture of the rear hull engine compartment of an M48A3 tank with its engine and transmission removed. Visible are the vehicle's internal fuel tanks. *Patton Museum*

Hitching a ride into combat on the turret of a U.S. Army M48A3 tank during the Vietnam War is a single American infantryman. Early-production M48A3 tanks as is the one pictured retain the small and cramped M1 cupola from the M48A2 and M48A2C tanks and the cross bar bracket on the front fender extensions. It retains the headlight clusters and rounded brush guards inherited from the M48A1 tank. *Patton Museum*

In April 1967, a commercial firm was awarded a contract to upgrade the U.S. Army's existing inventory of early-production M48A3 tanks. Among the many improvements made to the vehicle was the addition of a spacer ring under the existing M1 cupola that contained nine large periscopes as seen here. The upgraded M48A3 tanks were referred to as the M48A3 (Mod B) tanks. *TACOM*

A factory fresh early-production M48A3 (Mod B) poses for the company photographer. Notice the 2.2 kilowatt Xenon white light/infrared searchlight mounted just above the gun shield. The spacer ring under the M1 cupola was designated as the G305 turret riser. *Patton Museum*

The M1 cupola on the M48A3 (Mod B) tanks came with a new bulged hatch cover to provide the tank commander with more headroom as shown in this overhead picture. On the M48A3 (Mod B) tank the turret bustle rack, which was inherited from the M48A1 tank, was partly encased with a steel mesh screen to better retain smaller object stored there. *TACOM*

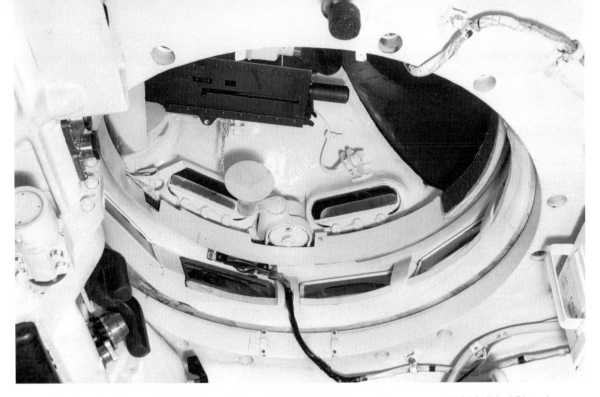

The interior layout of the M1 cupola fitted with the G305 turret riser on an M48A3 (Mod B) tank can be seen looking up from the tank commander's position. Notice that the .50 caliber (12.7mm) machine gun has been laid on its side to fit in the M1 cupola. *TACOM*

Visible in this interior picture of the tank commander's cupola on an M48A3 (Mod B) tank is the periscope sight M28C that the vehicle commander used to aim the .50 caliber (12.7mm) machine gun mounted in the M1 cupola. *Chun-Lun Hsu*

The four-man crew of a U.S. Marine Corps M48A3 tank nicknamed "The Grim Reaper" pose for the photographer in 1971. This tank had the G305 turret riser but had not yet gone through the Mod B upgrade program. The vehicle also retains the white light only 18-inch Crouse-Hinds searchlight that first appeared in use during the Korean War. *Ken Estes*

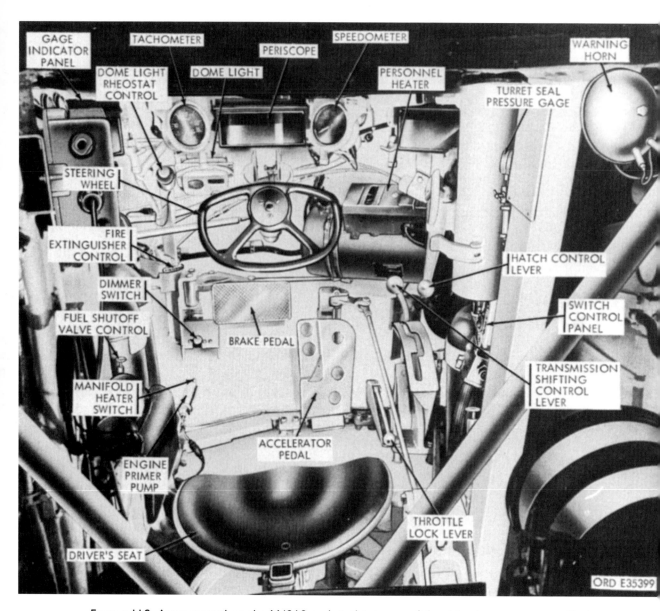

GAGE INDICATOR PANEL

TACHOMETER

DOME LIGHT RHEOSTAT CONTROL

DOME LIGHT

PERISCOPE

SPEEDOMETER

PERSONNEL HEATER

WARNING HORN

TURRET SEAL PRESSURE GAGE

STEERING WHEEL

FIRE EXTINGUISHER CONTROL

DIMMER SWITCH

FUEL SHUTOFF VALVE CONTROL

MANIFOLD HEATER SWITCH

ENGINE PRIMER PUMP

DRIVER'S SEAT

BRAKE PEDAL

ACCELERATOR PEDAL

THROTTLE LOCK LEVER

HATCH CONTROL LEVER

SWITCH CONTROL PANEL

TRANSMISSION SHIFTING CONTROL LEVER

ORD E35399

From a U.S. Army manual on the M48A3 tank is this image of the driver's compartment. Tank drivers were taught to position their vehicles as nearly level as possible as side slopes will cause the main gun to be canted and correction of range will be difficult for the gunner. *Patton Museum*

In this picture can be seen two late-production M48A3 (Mod B) tanks in service with the U.S. Marine Corps during a training exercise. Late-production M48A3 (Mod B) tanks can be readily identified by their identical headlight clusters and squared brush guards that first appeared on the M48A2 tank. *DOD*

Visible in this picture taken during the Vietnam War is a late production M48A3 (Mod B) tank. It has the new headlight clusters with the squared brush guards. The new squared front fender extensions have an embossed "X" stamped on them as can also be seen in this image. *Patton Museum*

Here we see a late-production Army of the Republic of South Vietnam (ARVN) M48A3 (Mod B) tank. Besides the identifying feature of the squared brush guards around the rear taillights, the vehicle pictured also has the embossed "X" pattern on the rear fenders not seen on early-production M48A3 (Mod) B tanks. *Patton Museum*

This is the original design for the brush guards around the rear taillights for the M48A2 through early production M48A3 (Mod B) tanks. *Michael Green*

The thick welded-on protruding steel bars that was affixed just above the two large steel louvered doors at the rear hull of late-production M48A3 (Mod B) tank is shown here. *Michael Green*

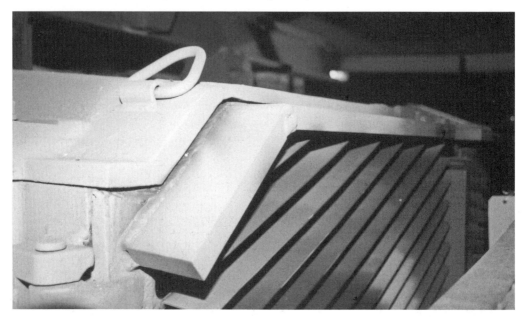

Chapter Five:

Patton Tanks
in the Vietnam War

In what the United States military and political senior leadership saw as a continuing struggle against the spread of Communist aggression, small numbers of American military advisors were sent to South Vietnam in 1956 to help train the South Vietnamese military to resist armed aggression sponsored by the Communist government of North Vietnam. An armed insurgency orchestrated by North Vietnam began in South Vietnam the following year. In response,

An U.S. Army M48A3 tank is pictured carrying a squad of American infantry into action somewhere in South Vietnam. The clean and uncluttered appearance of the vehicle identifies it as a recent arrival to the war. It would soon sport a vast array of miscellaneous external paraphernalia as the tankers learned to live for long periods of time out in the field. *Patton Museum*

The M48A3 tank pictured in South Vietnam bears all the hallmarks of a crew that has adapted to spending a long time in the field. Notice the non U.S. Army issued lawn chairs and the tarp over the top of the vehicle's turret to protect the crew from the hot sun. *Patton Museum*

additional American military advisors were sent to South Vietnam. By 1965, there were almost 20,000 American military advisors in South Vietnam.

As the South Vietnamese military floundered in its struggle against the Communist-sponsored insurgency, a decision was made in 1965 to officially commit American military ground forces to the growing conflict. The first American military ground force unit to arrive in South Vietnam was the 9th Marine Expeditionary Brigade (9TH MEB), which waded ashore near the coastal city of Da Nang, on 8 March 1965. The next day they brought ashore a platoon of five M48A3 Patton tanks belonging to the 3rd Tank Battalion. It took until July 1965 before the battalion's entire inventory of tanks was ashore in South Vietnam.

The first major combat encounter in South Vietnam that involved American tanks occurred in August 1965 when Marine Corps armor went up against enemy defensive positions in "Operation Starlite," and inflicted heavy losses on the Viet Cong troops manning them, with some losses to themselves.

The deployment of the Marine Corps 3rd Tank Battalion was against the wishes of the United States ambassador to South Vietnam and the Military Assistance Command, Vietnam (MACV) who felt that tanks were not suitable for

In this photograph we see an U.S. Army M48A3 tank in South Vietnam that has no doubt spent some time in the field as is evident by the broken and missing fenders. Despite many doubters the M48A3 tank showed that it could deal with the thickest of jungles in pursue of its missions. *Patton Museum*

the terrain, the counter-insurgency warfare taking place, and would place an undue strain on the logistical support system. With this thought in mind, U.S. Army ground force combat units that arrived in South Vietnam in 1965 had mostly been stripped of their tank battalions before arriving in country. At the same time, however, American units began encountering not just enemy guerilla units (referred to as the Viet Cong or VC) primarily using hit and run tactics, but main force North Vietnamese Army (NVA) units that often employed conventional tactics when the opportunity presented itself.

The appearance of main force NVA units in South Vietnam helped overcome the resistance of many senior leaders in the U.S. Army who initially felt that there was no place for tanks in the battle for Southeast Asia. U.S. Army ground force units that began arriving in South Vietnam in 1966 came with their full complement of M48A3 tanks. The successful combat use of these vehicles in various battles that took place between 1966 and 1967 convinced many of the non-believers that there was indeed a valuable place on the battlefields of Southeast Asia for tanks despite various terrain obstacles such as jungles and often less than optimum weather conditions.

The usefulness of the M48A3 tank and the other American armored fighting vehicles (AFVs) was proven again during the January/February 1968 surprise *Tet* offensive in which the VC and NVA moved out of their countryside lairs and struck at the urban population centers of South Vietnam and various large American military bases. It was American tanks and other mechanized vehicles that were able to respond to the enemy offensive in a timely manner with their mobility which disrupted many of the enemy's attack plans.

In the urban fighting that took place during the *Tet* offensive it was the firepower of the M114 90mm main guns mounted on the M48A3 tanks that often rooted out the VC and NVA from their urban strongholds. When the VC or NVA attempted to retreat from the cities of South Vietnam after their surprise

Pictured is an M48A3 tank embroiled in a South Vietnamese jungle. The elusive nature of the enemy during the Vietnam War meant that American ground combat units could expect contact with them at any time and from any direction. American tankers quickly learned to avoid establishing patterns of movement so the enemy could not plan their ambushes. *Patton Museum*

Plowing through a South Vietnamese jungle is an U.S. Army M48A3 tank. While the Viet Cong (VC) and North Vietnamese Army (NVA) would, as a general rule, tend to avoid contact with American units equipped with tanks however they would make effective use of the antitank weapons available to them when confronted. *Patton Museum*

offensive failed, it was American tanks that helped chase them back to their countryside base camps.

Sometime after the *Tet* offensive a shortage of M48A3 tanks caused the U.S. Army to ship an unknown number of M48A2C tanks to South Vietnam as is evident from pictorial evidence and the memories of American tankers who served during the conflict.

The only time American M48A3 tanks actually engaged in face to face combat with NVA armor during the Vietnam War occurred on the early evening of 3 March 1969, at the Ben Het Special Forces Camp, located in the mountains of South Vietnam's Central Highland. The NVA attacked the base with a number

American tanks like the U.S. Army M48A3 tanks pictured would most often encounter VC and NVA antitank weapons during ambushes. The VC and NVA would usually emplace their antitank weapons at each end of an ambush area where they could attack the first and last vehicles in their planned killing zone and also protect their flanks. *Patton Museum*

of Soviet-supplied PT-76 amphibious light tanks and at least one BTR-50 amphibious armored personnel carrier (APC). Unbeknownst to the NVA tankers there was a platoon of M48A3 tanks guarding the base. In the ensuing engagement the NVA lost two PT-76s and a BTR-50. The American tankers suffered two killed and slight damage to a single M48A3 tank.

American M48A3 tanks were employed in a variety of roles during their time in the Vietnam War; one of the best known was referred to as "search and destroy." These search and destroy operations were intended to locate enemy installations, attrite VC and NVA forces, and to destroy or remove the opponent's supplies and equipment. Less importance was given to seizing and holding critical terrain than to finding and finishing off the enemy. When enemy units were located they were attacked by a combination of maneuvering and blocking elements, both supported by artillery and aircraft. During many search and destroy operations, armor units (especially Cavalry elements) were initially engaged in area reconnaissance and intelligence gathering. When contact was made units they then undertook offensive operations, in what the American military describes as a "meeting engagement."

Many search and destroy operation began with dismounted infantry deployed as skirmishers to conduct a detailed hole-by-hole and bush-by-bush search with M48A3 tanks positioned well back. When a significant enemy combat unit was found, the dismounted infantry would seldom have the firepower needed to overcome the enemy at the point of contact. The tanks were then committed to the assault in order to destroy the enemy unit. A variation of this same mission would find the tanks leading the way and dismounted infantry following to protect the tanks from enemy antitank teams. In jungles and heavily overgrown areas, M48A3 tanks were employed to break pathways for the dismounted infantry, detonate antipersonnel mines, and support the dismounted infantry by firing on enemy defensive works and crew served weapons.

The combat engagement ranges that occurred during search and destroy operations between M48A3 tanks and their opponents were generally under 328

Besides placing antitank weapons at each end of an ambush area, the VC and NVA would also place them at intervals throughout the killing zone. Antipersonnel mines would be placed in trees to kill and injure personnel riding on the tops of tanks. Here we see an U.S. Army M48A3 tank crossing a South Vietnamese river. *Patton Museum*

Marines are pictured in the process of fueling an M48A3 tank. During ambushes of American tanks the VC and NVA would make sure that all their antitank weapons had designated alternate positions to protect the rear of the ambush force against encirclement by American reinforcements. *Patton Museum*

yards (300m); however, engagements at 16-27 yards (15-25m) were not uncommon, especially in areas of dense vegetation.

The most commonly employed main gun round M48A3 tanks used during search and destroy operations was called "canister." The canister round was designed solely for use against enemy personnel at relatively short range. When fired, the casing of the projectile split as the round left the muzzle of the tank's cannon releasing 1,281 steel pellets. The pellets moved down range in a conical pattern acting in the same manner as a shotgun blast. The canister round was capable of inflicting casualties in an area 33 yards (30m) wide and 202 yards (185m) deep. Its heavy use led to the improved "beehive" round for the M41 90mm main gun on the M48 series tanks designated the XM580E1 Anti-Personnel Tracer (APERS-T), which contained 4,100 small steel flechettes and was even more deadly against the VC and NVA.

Other types of operations American M48A3 tanks took part in during the Vietnam War included "clear and hold," which was aimed at driving enemy forces out of designated areas and keeping them out. There was also something

Pictured is an U.S. Army M48A3 tank that has shed it right track somewhere in South Vietnam. Due to terrain features or enemy mines American tank crews during the Vietnam War became extremely adept at quickly getting their vehicles up and running again for fear that a stationary tank would quickly become a prime target for the VC or NVA. *Patton Museum*

referred to as "security." Security operations for armor units included route security and convoy escort.

Route security required a large armored force for the entire time of a planned convoy mission. Bridges and other critical points first had to be secured with mobile tank-heavy outposts. There also were tank-heavy patrol actions, nick-named "roadrunner" operations, conducted at random times and in varying directions between the mobile outposts to fend off attempts by enemy forces to lay an ambush. To deter the enemy from emplacing mines during the hours of darkness, some M48A3 tank units conducted roadrunner operations at night, firing their tank's machine guns and canister main gun rounds to both sides of a planned convoy route at irregular intervals to disrupt any enemy plans at laying an ambush.

Convoy escort missions required a much smaller force of M48A3 tanks, and that only for the time needed to move the convoy from one point to another. In some cases, a combination of route security and convoy escort operations were employed to safely move convoys through enemy infested areas.

It was not an uncommon practice during the Vietnam War to open convoy routes each day by driving M48A3 tanks over them before permitting other vehicles to use the local road network. Due to the thick, elliptically shaped hull of the vehicle, it was normally able to absorb the shock of most mine explosions with only light to moderate damage. However, the VC and NVA would on occasion use much larger than normal charges as improvised mines that could cause extensive damage to M48A3 tanks and kill the crews that served upon them.

Mines were the most widely used weapon employed by VC and NVA forces against American and Allied tanks. Other tank killing weapons employed by the enemy included crew-served 57mm and 75mm recoilless rifles, as well as man-portable shoulder-launched rocket-propelled grenade launchers (RPGs). The VC and NVA originally used the Soviet-designed RPG-2 until it was in turn replaced by the longer ranged and more effective RPG-7, which fired a roughly five pound (2.25kg) rockets with a shaped-charged warhead out to 547 yards (500m) which could penetrate approximately 260mm of steel armor.

Growing frustration by the American public with the conflict in Southeast Asia and the continued losses suffered by the American military pushed the country's political and military senior leadership to begin withdrawing American troops

A U.S. Marine Corps M48A3 tank is seen carrying a group of Marine infantrymen. The vehicle is equipped with a Korean War vintage 18-inch Crouse-Hinds searchlight. Notice the infantrymen behind the tank are walking in the track impressions to avoid enemy antipersonnel mines or booby traps. *Patton Museum*

U.S. Army M48A3 tanks are shown jungle-busting in South Vietnam. This process caused a great deal of external damage to the vehicle as is seen with the missing fenders on the left side of the tank in the background. Unseen, is the wear and tear on the vehicle's suspension system and drive train. *Kent Hillhouse*

from Southeast Asia in 1969. The tempo of withdrawal increased in 1970 and 1971. The last remaining American military tank unit departed Southeast Asia in April 1972.

Filling the void left by American ground forces was the ARVN in a process known as "Vietnamization." Among the weapons left behind by the American military to aid the ARVN was a battalion's worth of M48A3 tanks, which were used to form the 20th Tank Regiment. The primary tank of the ARVN since 1965 had been the American-supplied M41A3 Walker Bulldog light tank, armed with a 76mm main gun.

With the departure of American ground forces from Southeast Asia, the emboldened military and political senior leadership of North Vietnam mounted a massive offensive operation against South Vietnam on the Easter weekend of 1972. In the vanguard of the NVA operation were up to 600 tanks, including Soviet-supplied T54 medium tanks armed with a 100mm main gun. However, American airpower assets in the region helped blunt the NVA offensive along with a surprisingly strong showing by many ARVN units. The AVRN 20th Tank Regiment took a heavy toll of NVA tanks during the spring invasion. During a single engagement along National Highway 9 on 9 April 1972, the M48A3 tanks of the ARVN 20th Tank Regiment destroyed sixteen T54 tanks and captured a Chinese-built copy of the T54 tank, designated the Type 59.

After being defeated in their 1972 spring invasion of South Vietnam, North Vietnamese leaders began rebuilding their ground forces in both North Vietnam and South Vietnam and awaited the departure of the American military from South Vietnam.

The turning point for the United States long involvement in the Vietnam War occurred when secret American negotiations to end the war in Southeast Asia finally bore fruit on January 23, 1973, and various combatants signed the Paris Peace Accords. America soon got back its last prisoners of war and slowly but surely withdrew the last vestiges of its military-power from the region. In response, the NVA mounted another massive tank-led offensive operation against South Vietnam on March 1, 1975. Without the cover of American airpower, the ARVN crumbled and Saigon, the capital of South Vietnam fell on 30 April 1975.

Pictured is a U.S. Army M48A3 tank somewhere in South Vietnam. The loader is armed with a Second World War-era M3 or M3A1 submachine gun, popularly known as the "Grease Gun." Notice the spare road-wheel attached to the rear turret bustle rack. *Patton Museum*

A standard tactics to fight off VC or NVA ambushes on routes from which vehicles cannot deploy was to form the M48A3 tanks into the herringbone formation seen here. The tanks rapidly close to within a few feet of each other to achieve a high density of firepower. *Patton Museum*

When escorting road-bound convoys made up of unarmored wheeled vehicles, M48A3 tanks would form a more spread-out herringbone formation as seen here in order to allow an open lane on the route to permit evacuation of the wounded, for resupply, or reinforcements. *Patton Museum*

The enemy's response to the deployment of large number of M48A3 tanks to South Vietnam was to increase the number of antitank mines they employed. The mines ranged from standard Eastern-Bloc pressure detonated devices up to massive improvised mines. A minesweeper team is shown checking a road for the waiting U.S. Army M48A3 tank. *Patton Museum*

This picture illustrates the damage a large improvised mine could do to an early-production U.S. Army M48A3 tank in South Vietnam. The improvised mines employed by the enemy included recovered dud American artillery shells and aerial-dropped bombs of varying sizes. *Patton Museum*

A U.S. Army M48A3 tank is pictured giving a ride to three infantrymen during the Vietnam War. American tankers learned that constant random variation of patrol actions was essential to keep the enemy off balance. When confronted by stronger forces, the VC or NVA would typically go into hiding or mass at pre-arranged rallying points. *Patton Museum*

M48A3 vehicle commanders in South Vietnam found the M1 cupola-mounted .50 caliber (12.7mm) machine gun difficult to load and operate in combat because of the very cramped nature and poor visibility of the cupola. Many decided to remove the weapon and mount it on top of the M1 cupola as seen here on a U.S. Marine Corps M48A3 tank nicknamed "Austrophodia." *Patton Museum*

During the Vietnam War, American tankers quickly learned that once you made a trail, you can assume that the enemy will mine it. The answer was to constantly cut new trails. Pictured is a U.S. Army M48A3 tank. Notice the tank commander is wearing a helicopter crewman's helmet. *Patton Museum*

A U.S. Army M48A3 tank is shown in South Vietnam loaded down with boxes of main gun ammunition after a helicopter resupply mission. Main gun rounds were transported in fiber containers packed two to a 131 pound wooden crate. *Kent Hillhouse*

This ghosted line illustration shows the ammunition storage arrangement in the M48A3 tank. The arrangement of munitions in the vehicle left it and others in the M48 tank series vulnerable to catastrophic destruction if a penetrating round struck the propellant in a main gun cartridge case. *Patton Museum*

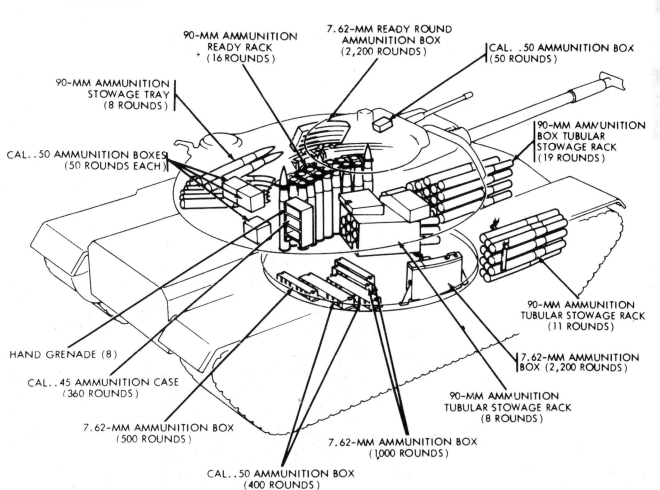

90-MM AMMUNITION READY RACK * (16 ROUNDS)

7.62-MM READY ROUND AMMUNITION BOX (2,200 ROUNDS)

CAL. .50 AMMUNITION BOX (50 ROUNDS)

90-MM AMMUNITION STOWAGE TRAY (8 ROUNDS)

90-MM AMMUNITION BOX TUBULAR STOWAGE RACK (19 ROUNDS)

CAL. .50 AMMUNITION BOXES (50 ROUNDS EACH)

90-MM AMMUNITION TUBULAR STOWAGE RACK (11 ROUNDS)

HAND GRENADE (8)

7.62-MM AMMUNITION BOX (2,200 ROUNDS)

CAL. .45 AMMUNITION CASE (360 ROUNDS)

7.62-MM AMMUNITION BOX (500 ROUNDS)

90-MM AMMUNITION TUBULAR STOWAGE RACK (8 ROUNDS)

7.62-MM AMMUNITION BOX (1,000 ROUNDS)

CAL. .50 AMMUNITION BOX (400 ROUNDS)

The vehicle commander of a M48A3 tank in South Vietnam poses for a friend's camera. Reflecting the often close-range nature of combat during the Vietnam War, he has by his side a Second World War vintage M1A1 Thompson .45 caliber submachine gun (minus the butt-stock) and a M1911A1 .45 caliber semi-automatic pistol. *Kent Hillhouse*

The U.S. Army's answer to the problems with the cramped M1 cupola and complaints about the poor level of visibility it offered was the introduction of the M48A3 (Mod B) tank with the G305 turret riser (fitted with nine periscopes) as is pictured here. Some of the periscopes show damage from bullet impacts. *Patton Museum*

Here we see an early production U.S. Army M48A3 (Mod) B tank with the G305 turret riser. This particular vehicle has become immobilized in a South Vietnamese rice field. Tank drivers are taught when confronted with a marshy area to keep their momentum up, follow a straight line, and avoid spinning the tracks. *Patton Museum*

Despite the U.S. Army's best intentions with the introduction of the G305 turret riser, tank commanders fighting in South Vietnam decided they still preferred to mount their onboard .50 caliber (12.7mm) machine gun on top of the M1 cupola as is pictured on this early-production M48A3 (Mod B) tank. The machine gun is mounted on a ground tripod and is being steadied with sandbags. *Patton Museum*

A close-up photograph of the mounting arrangement of a .50 caliber (12.7mm) machine gun on an M48A3 (Mod B) tank fitted with the G305 turret riser. The machine gun mount is welded to the front of the M28 periscope sight located in the roof of the M1 cupola that was used to aim and fire the weapon when mounted inside the cupola. *Patton Museum*

The vehicle commander of the M48A3 (Mod B) tank pictured had the M28 periscope sight located in the roof of the M1 cupola removed. In its place there is a flat metal plate bolted on top of the M1 cupola and attached to the plate is his .50 caliber (12.7mm) machine gun. At the ready, are several belts of linked .50 caliber (12.7mm) ammunition, with a typical combat mix of four rounds armor piercing incendiary (API), one round armor piercing incendiary-tracer (API-T). *Patton Museum*

Not all M48A3 tank commanders favored mounting their .50 caliber (12.7mm) machine guns on the roof of the M1 cupolas. The early production M48A3 (Mod B) tank being towed has a .30 caliber (7.62mm) machine gun mounted on the roof of the M1 cupola. Notice the early-type taillight brush guards on this vehicle. Patton Museum

Some M48A3 (Mod B) tank commanders decided to forego mounting their .50 caliber (12.7mm) machine gun on top of their M1 cupola and instead mounted it at the loader's position as seen in this photograph. Some tank crews mounted an additional machine gun at the rear of their turrets. *Patton Museum*

Here we see an M48A3 (Mod B) tank in the service of the Army of the Republic of Vietnam (ARVN) that has been modified to have the vehicle's .50 caliber (12.7mm) machine gun operated from the loader's position. *Patton Museum*

The most feared antitank weapon employed by the VC and NVA was the Soviet-designed RPG-7. It was a shoulder-fired weapon firing a shaped charged antitank grenade that could penetrate the thickest armor on the M48A3 tank. An American soldier in an opposing forces uniform is seen here posing with RPG-7. *DOD*

Pictured is an RPG-7 penetration of the thick armor of the gun-shield on an M48A3 tank in South Vietnam, which injured everybody in the vehicle. Despite having a maximum effective range of several hundred yards, the VC and NVA felt the closer the better and typically fired them at ranges of 55 yards (50m) and under to improve accuracy. *Kent Hillhouse*

In this picture we see an RPG-7 projectile that bounced off a return roller on an M48A3 tank in South Vietnam and then failed to penetrate the lower hull side of the vehicle. The shaped charge grenades fired from the RPG-7s were stabilized in flight by fins located at their rear. The gash in the lower hull side makes it look like a penetration. *Kent Hillhouse*

The VC and NVA favored attacking American military ground forces at night to offset their firepower advantage. It was a common practice to install cyclone fencing (attached to engineer stakes) in front and around M48A3 tanks when halted at night to pre-detonate incoming RPG-7 projectiles. Pictured is an example of that practice. *Patton Museum*

This picture was taken from the top of an U.S. Army M48A3 (Mod B) tank in South Vietnam. The tank in the foreground carries a large bundle of cyclone fencing material wrapped around a number of engineering stakes. The box-like brush guard around the rear tail lights of the tank in the foreground mark it as a late-production M48A3 (Mod B) tank. *Patton Museum*

Since erecting a cyclone fence around a moving tank was impractical, the crew of this early-production U.S. Army M48A3 (Mod B) tank built themselves a sandbag fort on the top of their vehicle's turret to survive attacks by enemy RPG-7s. The additional weight of the sandbags increased the wear and tear on the drivetrain. *Patton Museum*

Late-production M48A3 (Mod B) tanks can normally be identified by the new squared brush guards around the front hull-mounted headlight clusters. Those squared brush guards can be seen in this picture showing a couple of tankers loading up their vehicle with main gun rounds in South Vietnam. *Patton Museum*

Here we see a number of late-production M48A3 (Mod B) tanks identified by their square brush guards around the rear tail lights on a firing range in South Vietnam. The rear engine decks of the vehicles shown are covered by a wide variety of gear. *Patton Museum*

Running at speed in South Vietnam is this late-production U.S. Army M48A3 (Mod B) tank identified by its squared brush guards around the front hull headlight clusters. Barely visible is the barrel of a 7.62mm M60 machine gun mounted at the loader's position. During convoys it was typical to have one armored vehicle per five to ten wheeled vehicles. *Patton Museum*

Here we see two U.S. Army M48A3 tanks in South Vietnam without the Mod B turret risers but with late production Mod B front-hull mounted headlight clusters with squared brush guards. Notice that the front hull mounted fender extensions on both tanks have either been cut off or torn off due to dense vegetation. *Patton Museum*

Working its way through a South Vietnamese jungle is this U.S. Army late-production M48A3 (Mod B) tank. Enemy tactics during the Vietnam War relied on speed, secrecy, surprise and deception for their success. *Patton Museum*

A U.S. Marine Corps early-production M48A3 (Mod B) tank crosses over a rice paddy dike some-where in South Vietnam. The large wheel at the front of the tank's suspension system is one of two front hull compensating idlers that helped to eliminate slack in the vehicle's two sets of tracks under dynamic conditions. Slack is the stretching of a tank's tracks, which can then disengage them-selves from the vehicle's suspension system if not corrected for. *Patton Museum*

The U.S. Army crew of this late-production M48A3 (Mod B) tank in South Vietnam have taken the time to decorate the front hull of their vehicle with a little field artwork. During the Vietnam War the enemy prepared extensive defensive systems throughout their areas of operation, which often took American troops an inordinate amount of time to subdue. *Patton Museum*

In the close confines of the jungle terrain of South Vietnam the enemy would get as close to American tanks as possible to take advantage of the dead spaces around the vehicles where their weapons could not be sufficiently depressed to engage them. Somewhere in South Vietnam is this early-production M48A3 (Mod B) tank surrounded by the local vegetation. *Patton Museum*

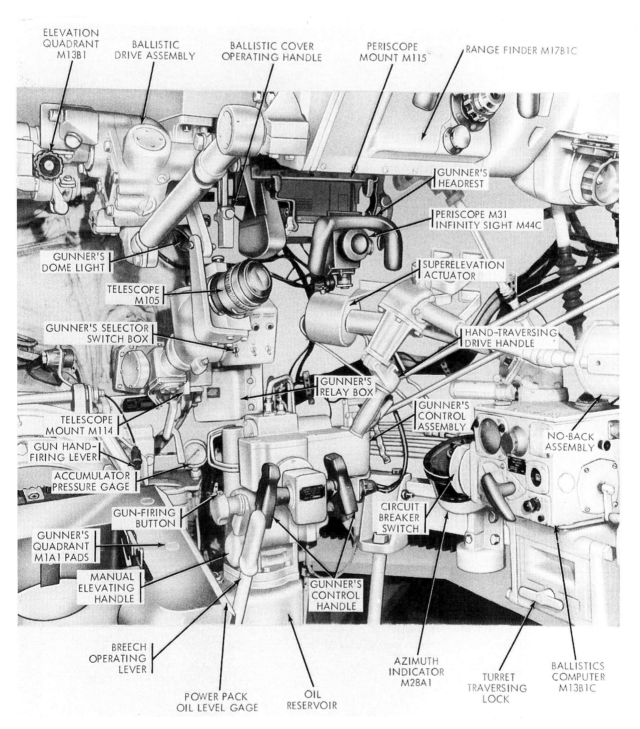

ELEVATION QUADRANT M13B1

BALLISTIC DRIVE ASSEMBLY

BALLISTIC COVER OPERATING HANDLE

PERISCOPE MOUNT M115

RANGE FINDER M17B1C

GUNNER'S HEADREST

PERISCOPE M31 INFINITY SIGHT M44C

GUNNER'S DOME LIGHT

TELESCOPE M105

SUPERELEVATION ACTUATOR

GUNNER'S SELECTOR SWITCH BOX

HAND-TRAVERSING DRIVE HANDLE

GUNNER'S RELAY BOX

GUNNER'S CONTROL ASSEMBLY

NO-BACK ASSEMBLY

TELESCOPE MOUNT M114

GUN HAND-FIRING LEVER

ACCUMULATOR PRESSURE GAGE

GUN-FIRING BUTTON

CIRCUIT BREAKER SWITCH

GUNNER'S QUADRANT M1A1 PADS

MANUAL ELEVATING HANDLE

GUNNER'S CONTROL HANDLE

BREECH OPERATING LEVER

POWER PACK OIL LEVEL GAGE

OIL RESERVOIR

AZIMUTH INDICATOR M28A1

TURRET TRAVERSING LOCK

BALLISTICS COMPUTER M13B1C

From an U.S. manual comes this picture of the gunner's position on an M48A3 tank with all the various devices and components that surround him. The gunner could aim the main gun or the coaxial machine gun with either a magnified telescope sight or a magnified periscope sight. *Patton Museum*

Here we see two late production U.S. Army M48A3 (Mod B) tanks enmeshed in a South Vietnamese jungle. Under such condition the enemy would attempt to throw hand grenades into the normally open hatches of the American tanks. It was not unheard of for enemy personnel to actually attempt to mount tanks to engage their crews at point-blank range with their small arms or deposit satchel charges. *Patton Museum*

Reflecting the often close-range fighting that occurred during the Vietnam War between American tanks and enemy troops, the crews of the two early-production U.S. Army M48A3 (Mod B) tanks shown have strapped E8 tear-gas dispensers onto the tops of their searchlights. *Patton Museum*

Pictured is a formation of late production U.S. Army M48A3 (Mod B) tanks in South Vietnam. The tank crews have affixed engineer stakes horizontally over the lower portion of the vehicle's glacis to prevent vegetation riding up and injuring the driver when jungle busting. *Patton Museum*

An early-production U.S. Marine Corps M48A3 (Mod B) tank is shown taking part in the recapture of the ancient Vietnamese Imperial City of *Hue* in February 1968. In an urban environment such as this, the preferred main gun round would be HE with a delayed action fuze set to detonate after penetrating the walls of an enemy-held building. *Patton Museum*

A dismounted U.S. Army tanker is guiding the driver of his early-production M48A3 (Mod B) tank over a new track so it can be connected and the vehicle placed back in service. Convoy protection duties by tanks took a heavy toll of their tracks. *Patton Museum*

Army of the Republic of South Vietnam (ARVN) troops ride into action on a U.S. Army M48A2C tank. The most noticeable features that identifies the vehicle pictured as an M48A2C tank is the lack of a large box-like dry type cleaner box as seen on M48A3 tanks and the lack of track tension idler between the last set of road wheels and the rear hull-mounted drive sprocket. A U.S. Army combat cameraman is standing next to the vehicle. *Patton Museum*

Chapter Six:

M48A5 Patton Tank

Designed from the onset to have the ability to mount a 105mm main gun, the M48 tank series did not feature that weapon on a series production vehicle until the advent of the 105mm gun full tracked combat tank M48A5. What finally pushed the U.S. Army to up-gun a portion of its M48 series tank inventory with the M68 105mm main gun was the October 1973 Arab-Israeli War, also referred to as the "Yom Kipper War." During that conflict, the United States Government sided with the country of Israel and supplied it with a large number of M60A1 tanks from U.S. Army war reserve stocks. Lacking the industrial base to quickly replace those M60A1 tanks, the M48A5 tank was seen a quick fix to replenish the U.S. Army's inventory.

The first batch of M48A5 tanks rolled off the conversion line of the Anniston Army Depot (ANAD) between October 1975 and December 1976. A second

Pictured is an Israel Defence Forces (IDF) M60A1 tank knocked out during the fierce fighting that marked the early stages of the 1973 Yom Kippur War. The IDF lost so many tanks in combat to the armies of its Arab neighbors in that conflict that the U.S. Government stripped its own war reserve stocks of tanks to re-equip the IDF. *IDF*

To help replace the many tanks supplied to the IDF during and after the 1973 Yom Kippur War, the U.S. Army set about upgrading a portion of its inventory of M48A3 tanks with 105mm main guns and other improvements. The upgraded M48A3 tank became the M48A5 tank, an example of which is shown here. *TACOM*

batch of M48A5 tanks came out of Anniston between October 1976 and March 1978. The third and final batch of M48A5 tanks was completed by ANAD in December 1979. In total, ANAD would convert over 2,000 early-model Patton tanks into the M48A5 configuration. The M48A5 tank weighed in at 108,000lbs (49mt) and had a cruising range of about 300 miles (483km).

The actual conversion process by ANAD between the three different batches of M48A5 Patton tanks varied a great deal in the complexity of work required. The initial batch of 501 vehicles consisted of M48A3 tanks being converted to the M48A5 standard. About 400 of these vehicles were M48A3 tanks returned to the U.S. Army by the U.S. Marine Corps, which had re-equipped themselves with M60A1 tanks between 1974 and 1976.

The process required to bring the M48A3 tank up to the M48A5 tank standard required only eleven new major sub-assemblies as the M48A3 tank had already been upgraded with as many M60 series tank components as possible

This picture shows a rear view of an early production M48A5 tank with the M48A3 (Mod B) tank configuration features, which included the bulged tank commander's overhead hatch. Notice the late production M48A3 (Mod B) tank configuration strengthened rear fender extensions with the embossed X-pattern. *TACOM*

during its original conversion process done back in the early 1960s. The second and third tranches of early-model M48 series tanks consisting of 708 and 960 vehicles respectively involved converting M48A1 tank into the M48A5 configuration, using sixty-seven new major sub-assemblies.

The initial examples of the M48A5 tanks that came out of ANAD did not differ much from the M48A3 tank that they had been converted from, except for the addition of the M68 105mm main gun. They retained the modified M1 cupola with the bulged hatch cover and the G05 turret cupola riser. Also seen on the early conversion units of the M48A5 tank was the turret interrupting bar mounted in front of the M1 cupola. However, even before the first M48A5 tank was shipped out of ANAD, the U.S. Army decided to incorporate a number of improvements to the vehicle based on Israel Defense Force's (IDF) combat experience with various versions of the M48 series tanks gained during the October 1973 Arab-Israeli War.

The Israeli-based improvements to the M48A5 tank first began appearing on the conversion line at ANAD in August 1976. They included an increase in 105mm main gun ammunition storage from forty-three to fifty-four rounds. External changes included the replacement of the unsatisfactory M1 cupola with

The bulk of the U.S. Army's inventory of M48A5 tanks served with U.S. Army National Guard and U.S. Army Reserve units up till the 1990s. The vehicle pictured belonged to the California National Guard and is equipped with the G305 turret riser it inherited from a M48A3 (Mod B) tank. *Michael Green*

an Israeli-designed low-profile cupola that had three periscopes and was fitted with a two-position scissors-type mount for the vehicle commander's 7.62mm M60D machine gun. The vehicle's commander's overhead hatch on the Israeli-designed low-profile cupola could be locked into a fully opened vertical position, or if he was worried about threats from above (such as small arms fire or artillery air bursts,) he could raise his overhead hatch to an horizontal position that lay just a few inches above the rim of the cupola. Vehicles with the Israeli-inspired upgrades were designated the M48A5PI; the letters "PI" stood for product improved. Eventually, all the early production M48A5 tanks with the M48A3 (Mod B) features were brought up to the M48A5PI standard and the "PI" designation was dropped.

The majority of M48A5 tanks went directly into service with the U.S. Army Reserve and U.S. Army National Guard tank units. However, 140 of them were assigned to the U.S. Army's 2nd Infantry Division stationed in South Korea so that their M60A1 tanks could be returned to the United States for overhaul. The first of the M48A5 tanks arrived in South Korea in June 1978. As the M48A5 tank series finally passed from U.S. Army service in the 1990s, many went to friendly foreign armies under the Military Assistance Program (MAP).

Besides the U.S. Army's plans to arm its M48A3 tanks with 105mm main guns, which never came to fruition, there was another project that, if implemented, would have resulted in the M48 series tanks featuring a 105mm main gun.

In the early 1960s the U.S. Army envisioned placing a newly-designed turret mounting a combination 152mm gun/missile launcher system on the chassis of the original M60 tank. The mating of these two components would have resulted in the fielding of a vehicle with the official designation 152mm gun full tracked combat tank M60A2. It was anticipated that many redundant M60 tank turrets, armed with a 105mm main gun, would then become available for other uses. Early plans called for mounting 243 of these redundant M60 tank turrets on the chassis of M48A1 tanks upgraded to the M48A3 configuration.

As events unfolded, the number of M60A2 tanks was cut back and the proposed M48 series tank with the M60 tank turret was cancelled. If the program had gone forth, the standardized vehicle would have been designated as the 105mm gun full tracked combat tank M48A4. Although up-gunned M48 series tanks armed with an M68 105mm main gun in Israeli service were unofficially referred to as the M48A4 tank in some circles, it was never an official U.S. Army designation.

Taken inside a California National Guard M48A5 tank is this view from the loader's position of the tank commander looking through his coincidence rangefinder while the gunner observes the target through his own sighting system. It is normally the job of the tank commander to give the fire commands. *James Edwards*

Pictured are California National Guard M48A5 tanks during a training exercise. Notice the vehicle in the foreground has the original M1 cupola but no G305 turret riser, which meant it would have started life as an M48A1 tank before being converted into an M48A5 tank. *Michael Green*

Pictured is an M48A5PI with various Israeli-inspired upgrades, including the new low-profile cupola and the addition of two 7.62mm M60D machine guns located at both the tank commander and loader stations. *Patton Museum*

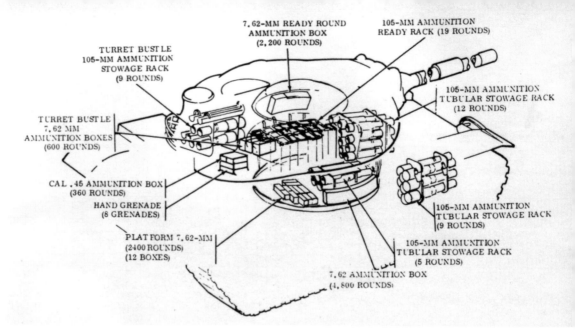

7.62-MM READY ROUND
AMMUNITION BOX
(2,200 ROUNDS)

105-MM AMMUNITION
READY RACK (19 ROUNDS)

TURRET BUSTLE
105-MM AMMUNITION
STOWAGE RACK
(9 ROUNDS)

105-MM AMMUNITION
TUBULAR STOWAGE RACK
(12 ROUNDS)

TURRET BUSTLE
7.62 MM
AMMUNITION BOXES
(600 ROUNDS)

CAL .45 AMMUNITION BOX
(360 ROUNDS)

HAND GRENADE
(8 GRENADES)

105-MM AMMUNITION
TUBULAR STOWAGE RACK
(9 ROUNDS)

PLATFORM 7.62-MM
(2400 ROUNDS)
(12 BOXES)

105-MM AMMUNITION
TUBULAR STOWAGE RACK
(5 ROUNDS)

7.62 AMMUNITION BOX
(4,800 ROUNDS)

This line illustration shows the stowage arrangements for the 105mm main gun rounds in the M48A5PI tank. It also shows the stowage location of the machine gun and small arms ammunition, and the hand grenades carried inside an M48A5PI tank. *Patton Museum*

A side view of an M48A5PI tank. An important spotting feature of the M48A5PI tank is the M68 105mm main gun with a bore evacuator located roughly midway on the barrel. The M68 105mm gun on the M48A5-series tank is a modified license-built copy of the Royal Ordnance L7 series 105mm tank gun. *Patton Museum*

A California Army National Guard M48A5PI tank is shown. The 7.62mm M60D machine guns are not fitted on this vehicle. The thickest armor was on the front of the vehicle's turret and was 178mm, while the front upper hull armor was 110mm thick, sloped at 60 degrees. *Michael Green*

This picture shows the tank commander and loader of an M48A5PI tank operating their 7.62mm M60D machine guns. One can imagine the great deal of restraint that would have to be shown by them when firing their machine guns in the heat of combat and not hitting their searchlight. *Patton Museum*

Pictured are two California Army National Guard M48A5PI tanks in hull defilade positions. Typically a tank commander would position his tank in turret defilade when not engaging targets and only return it to hull defilade position when ready to engage a target. The vehicle's M60D machine guns are equipped with blank firing adapters since this is a training exercise. *DOD*

The Continental 12-cylinder supercharged AVDS-1790-2D air-cooled diesel engine and transmission have been removed from the California Army National Guard M48A5PI tank pictured. Top speed of the M48A5PI tank on a level road was 30mph (48kph) and it had a cruising range of about 300 miles (483km). *Michael Green*

The strange looking device located just behind the bore evacuators of the M48A5PI tank pictured is a pyrotechnic device that simulates the firing of the main gun by emitting a large puff of smoke. The strap with several small square sensors around the tank's turret is the Multiple Integrated Laser Engagement System (MILES) used to determine hits during training. The MILES is integrated with the yellow light at the left rear of the vehicle which would strobe to signal a lethal hit. *DOD*

Chapter Seven:

Patton Tanks in Foreign Service

Ever fearful of Soviet military aggression towards Western Europe, American President Harry S. Truman signed the Mutual Defense Assistance Act on 6 October 1949. An important component of that act was the Military Assistance Program (MAP), which would provide billions of dollars' worth of American military equipment to the countries that made up the North Atlantic Treaty Organization (NATO) formed on 4 April 1949.

It was intended that the American military equipment supplied under the MAP would help expedite the rebuilding of various Western European armies

On display at the Royal Museum of the Army and Military History located in Brussels, Belgium, is this M46 tank in the markings of the Belgium Army. This particular display vehicle is missing the assistant driver/bow gunner's hatch, which has been plated over. *Pierre-Olivier B*

Belonging to the Israeli Armored Corps Memorial Site and Museum in Latrun, Israel, is this M47 tank captured from the Royal Jordanian Army Corps during the 1967 Six-Day War. The vehicle is no longer in its original paint scheme. It has the late production T-shaped blast defector. *Vladimir Yakubov*

decimated during the Second World War, and in turn act as a key deterrent to any Soviet military designs in that part of world. As time went on, the United States would extend its MAP to many countries outside of NATO that were perceived as being friendly to American worldwide military and economic interests.

Among the many types of American military aid extended to countries around the world under MAP were relatively large numbers of surplus American tanks. Reflecting the limited number of M26 series tanks converted into M46 and M46A1 series tanks (1,170) only a small number were ever supplied under MAP to Belgium, France, and Italy. The M46 and M46A1 series tanks would only last in NATO service for a short period of time before large numbers of M47 tanks became available for use in the 1950s.

M47 Tank

Of the 8,576 American M47 tanks built, the majority spent their service lives with foreign armies. NATO armies were the largest recipients of the tank under MAP. At one point in time, the Italian Army boasted an inventory of over

The French Army inventory once contained a large number of M47 tanks. A small number of these tanks have been preserved as is the example pictured. The M47 tank would remain in French Army service until replaced by the French-designed and built AMX30 tank in the late 1960s. *Christopher Vallier*

2,000 M47 tanks all of which have long been retired from service.

The West German Army took into service during the 1950s over 1,000 M47 tanks with the French and Belgium Armies taking on about 800 M47 tanks each. It was under French colors that the M47 tank first entered into a shooting conflict during the short-lived Suez Crisis of 1956. The French M47 tanks were retired in the 1960s as were those of the West German and Belgium armies.

Austria took into service over 150 M47 tanks in the 1950s, which remained in service until the 1980s. Another European country that took the M47 tank into service was Spain, which also received about 400 M47 tanks under MAP in the 1950s. Most of these were eventually modernized and designated the M47E. There were addition upgrades to these tanks resulting in the M47E1 and M47E2 designations. None of the upgraded M47 tanks currently remain in service with the Spanish Army.

It was the M47 tanks of the Spanish Army that portrayed German Tiger tanks in the 1965 Hollywood-made movie "Battle of the Bulge," starring well-known American actor Henry Fonda, and the 1970 Hollywood-made movie "Patton," starring actor George C. Scott playing the famous American, General George S. Patton.

Under MAP the Turkish Army took into service over 1,300 M47 tanks. In the 1974 Turkish invasion of Cyprus, their American-supplied M47 tanks saw some limited action against Cypriot T34/85 medium tanks. The Turkish Army retired its fleet of M47 tanks in the 1960s.

Greece, like its neighbor Turkey, was provided with over 300 M47 tanks, beginning in the 1950s, all of which have long since been retired from service.

Iran received a fleet of roughly 400 American-supplied M47 tanks in the late 1950s, when it was under the control of a pro-U.S. Government. Many of these same tanks would later see service during Iran's war with Iraq that lasted from 1980 to 1988. The Iranian M47 tanks that would go into combat against Iraqi ground forces were an upgraded version of the vehicle designated the M47M.

Missing the brush guards around the front hull mounted headlight clusters and its bow machine gun, is this preserved former French Army M47 tank. The elongated and angular turret design of the vehicle was considered to offer a much higher level of ballistic protection for the turret than found on the M26 or M46 series tanks. *Pierre-Olivier B.*

This picture shows the rear turret and hull of a former French Army M47 tank now preserved as a memorial. The wire mesh panel attached to the bottom of the tank's rear hull plate could be lowered to a horizontal position so infantrymen could stand on it. To protect the infantrymen riding on the rear of the tank, the muffler exhaust pipes are aimed skyward and there are metal flaps to deflect engine heat upwards. *Pierre-Olivier B.*

These tanks had been modernized in Iran by the American firm of Bowen-McLaughlin Company (BMY) between February 1972 and March 1974.

The Iranian M47 tanks upgraded by BMY to the M47M configuration featured an improved fire-control system and the Continental AVDS-1790-2A air-cooled diesel engine from the American M60 series of tanks. Eliminated on the M47M was the bow-gunner's position. In his place went a rack for the storage of twenty-two additional rounds of 90mm main gun ammunition.

In the 1977-1978 conflict between Somalia and Ethiopia, the Ethiopian Army lost a number of American-supplied M47 tanks in combat. During the American military involvement in the country of Somalia between 1992 and 1994, there were at least two incidents where American military forces engaged in combat with Somalia-operated M47 tanks.

Other countries that have at some point in time had the M47 tank in service include South Korea, Pakistan, Jordan, Portugal, Taiwan, and the former country of Yugoslavia.

M48 Series Tanks

The West German Army took into service roughly 200 M48A1 tanks, over 1,000 M48A2 tanks and almost 500 M48A2C tanks. The West German armament industry would modernize 650 M48A2 and M48A2C tanks between 1978 and 1980. Among the many upgrades to the vehicle were a 105mm main gun and a white light/infrared searchlight. These modernized vehicles were designated the M48A2GA2. There are no longer any M48 series tanks in German Army service as they were either scrapped in 1992 and 1993, or given to the Turkish Army.

Turkey began receiving American-supplied M48 series tanks under the MAP, as well as those deemed excess to the West German Army needs, starting in the 1960s. In total, they received about 3,000 early-model M48 series tanks. Beginning in the early 1980s and ending at the start of the 21st Century, the Turkish Army embarked on an ambitious program to upgrade the majority of its inventory of early-model M48 series tanks to the American M48A5 tank configuration. In Turkish Army service the modernized vehicles were designated as the M48A5T1. Subsequent upgrades have resulted in an M48A5T2 version in Turkish Army service, which remains in use today.

A restored M47 tank belonging to the collection of the German Army tank museum is shown during a demonstration run. The American-supplied M47 tank would form the armored backbone of the new West German Army in the 1950s and provide their tank crews with basic tactical skills. *Thomas Anderson*

Posing for the company photographer is this brand-new M47M tank in the markings of the U.S. Army despite the fact that it would never see service with the American military. Notice the infrared suppression rear engine deck that first appeared on the M48A2 tank. The embossed X-shaped pattern on the rear fenders and the square brush guards around the rear hull taillights come from the late-production M48A3 (Mod B) tank. *TACOM*

The Greek Army eventually upgraded the bulk of its fleet of 800 early-model M48 series tank to either the American M48A3 or M48A5 configuration. Only those M48 series tanks upgraded to the M48A5 configuration currently remain in Greek Army service.

Spain also received American-supplied M48A3 and M48A5 tanks that were cycled through various modernization processes over the years. There are no M48 series tanks currently remaining in Spanish Army service.

Pakistan's Army employed both M47 tanks and early model M48 series tanks during a brief war that broke out between their country and India, in September 1965. The American-supplied tanks used by the Pakistani Army were frittered away in pointless attacks against well-laid out India Army defensive positions resulting in a loss of a great many of their tanks.

As the most continuously pro-Western Arab nation in the Middle East, Jordan has long been the recipient of a great deal of American military aid over the decades. In the 1960's, the Jordanian Army received roughly 200 early-model

When the opportunity arose to acquire the more advanced M48 series tanks, the West German Army quickly divested itself of its inventory of M47 tanks. Pictured is a West German Army M48A1 tank during a training exercise. *Patton Museum*

M48 series tanks. Many of these tanks would be lost in combat to Israeli airpower during the 1967 conflict between Israel and its Arab neighbors, best known as the "Six-Day War." The United States Government quickly made good the Jordanian losses with another batch of M48 series tanks. These were eventually superseded in Jordanian military service by American-supplied M60 series tanks.

The largest user of the M48 series tanks in combat was the Israeli Army. Israel had first expressed interest in acquiring the tanks from the United States in early 1964. The Israeli request was denied since the U.S. Government was loath at that point in time to antagonize Israel's various Arab neighbors. Instead, the Israeli Government was told to approach the West German Government. The West German Army had recently decided to pull out of service its inventory of M48A1 tanks and the West German Government therefore approved the transfer of 150 of those tanks to Israel in a secret deal organized in July 1964. However, press leaks about the deal caused a political outrage in Germany and in the Middle East. As a result, no more than forty M48A1 tanks made it to Israel before the deal was canceled.

The U.S. Government became much more receptive to the Israeli request for the M48 series tank in 1965. This change of heart came about due to the Soviet Union supplying large numbers of tanks to the Egyptian Army. To assist Israel in maintaining a certain level of military parity with its hostile neighbors, in 1965 the U.S. Government approved a one-time sale to the state of Israel of roughly 200 M48 series tanks almost evenly divided between M48A1 and M48A2 versions. The deal also included spare parts and up-arming conversion kits.

The Israeli armament industry began up-gunning the IDF's inventory of roughly 250 early-model M48 series tanks in 1966 with M68 105mm main guns and Continental AVDS-1790 diesel engines. This tank up-grading process was far from complete by the time the 1967 Six-Day War began, and it is reported that only a single company of twenty Israeli M48 series tanks boasted the more powerful 105mm main guns and the new diesel engines.

The majority of M48 series tanks that saw service with the Israeli Army during the 1967 Six-Day War retained their original 90mm main guns and gasoline engines. Despite this, they did extremely well in combat against Egyptian Army

A West German Army M48A2 tank is pictured on a training exercise. Both the .50 caliber (12.7mm) machine gun mounted in the M1 cupola and the blast deflectors have been covered with canvas tarps to protect them from the elements. *Patton Museum*

Forming part of the collection of the German Army tank museum located at Munster, Germany, is this operational M48A2C tank. The box-like device mounted on a bracket attached to the tank's turret is an IR/white light searchlight. *Thomas Anderson*

units equipped with Soviet-supplied T34/85s medium tanks and IS-3M "Stalin" heavy tanks. The Israeli Army did not use their M48 series tanks in their fight with the Jordanian Army during the Six-Day War for fear that there might be friendly fire incidents when both sides were employing the same type of tank in battle.

In Israeli Army service the M48A1 tank was referred to as a *Magach* 1, while the M48A2 Patton tank bore the designation *Magach* 2. Early model M48 series Patton tanks upgraded with 105mm main guns and a diesel engine were assigned the designation *Magach* 3. The English translation for *Magach* is "battering ram."

Following the 1967 Six-Day War both the Soviet Union and the United States rebuilt the inventories of their respective client states. The United States provided Israel with additional M48 series tanks as well as M60 series tanks.

Based on lessons learned during the fighting that took place during the 1967 Six-Day War, the Israeli Army decided to dispense with the M1 cupola on their M48 series tanks. It was believed that the added height of the M1 cupola was a battlefield weakness that left the tank commander poorly protected from enemy fire. Like their American counterparts, Israeli tank commanders also felt that the M1 cupola was both extremely cramped and offered very poor visibility when the overhead hatch was closed. In its place the IDF Armored Corps installed a low-profile machine gun armed cupola designed and built by the Israeli civilian firm of *Urdan*.

On a demonstration run is an operational M48A2C tank belonging to the German Army tank museum. The first M48 series tanks were delivered to the West German Army in 1958. The German tankers considered them much more reliable than the M47 tank. *Thomas Anderson*

Israeli tank crews also found out the hard way during the Six-Day War that the fluid used in the constant pressure hydraulic system employed to traverse the turret and elevate and depress the main guns on the M48 series tanks was extremely flammable. When the lines were punctured, they would spray the high pressure fluid all around the interior of the vehicle's crew compartments, creating an extreme fire hazard. A partial solution was to employ a far less flammable hydraulic fluid.

By the time the 1973 "Yom Kippur War" broke out, the bulk of the IDF's inventory of roughly 800 M48 series tanks had been both up-gunned with M68 105mm main guns and had their gasoline engines replaced with American-supplied diesel engines. Due to the steep losses incurred by Israeli M48 series tanks during the opening stages of the Yom Kippur War against Egypt, the United States began another a resupply effort that included more M48 series tanks and M60 series tanks. The American-supplied M60 series tanks would replace the last of the M48 series tanks in IDF service by the mid-1980s.

It has been reported that the South Korean Army currently has an inventory of 800 M48 series tanks, with half having been upgraded to the M48A5 configuration with 105mm main guns and diesel engines. The most recent upgrade to the South Korean Army M48A5 tank fleet includes armored side skirts and an improved fire control system. The designation for the upgraded tank is M48A5K.

An interesting variation on the M48 series tanks occurred in the late 1980s when Taiwan's Army installed a 105mm main gun in the turrets of their early-model M48 series tanks and then mounted them on the chassis of M60A3 tanks. Incorporating an advanced fire-control system, the vehicle is referred to as the CM-11. The same upgraded turret armed with a 105mm main gun and advance fire-control system mounted on a slightly modified M48A1 chassis is referred to as the CM-12.

On display at the German Army tank museum is what the West German Army referred to as the M48A2GA2, which was an upgraded M48A2C tank armed with a 105mm gun. It also featured a new square gun shield, as well as a new vehicle commander's cupola. *Frank Schulz*

Pictured are two West German Army M48-series tanks in an open air shed. The one on the left is an M48A2C tank and one on the right is an upgraded M48A2GA2 tank. Clearly visible is the much larger and thicker gun shield on the M48A2GA2 tank and the mid-position bore evacuator of its 105mm main gun. *Michael Green*

A convoy of South Korean Army M48A3 tanks passes by the photographer during a training exercise. The vehicles are fitted with an IR/white light searchlight. It is reported that the South Korean Army M48A3 tank inventory has been fitted with an indigenously-designed fire-control system. Vehicles so modified are designated as the M48A3K tank. *DOD*

This picture is of a heavily camouflaged South Korean Army M48A5PI tank. Notice that the two 7.62mm M60D machine guns and their mounts are missing. In their place is a single .50 caliber (12.7mm) machine gun on a different type of mount. This vehicle is also fitted with a smoke grenade launcher system. *DOD*

Out for a test drive is this Spanish Army M48A5E2 tank fitted with an IR/white light searchlight. As with many other countries who received American-supplied M48 series tanks, the Spanish Army upgraded them with more advanced fire control systems to keep them on par with the tanks of other nations. *DOD*

The majority of IDF M48 series tanks that saw combat during the 1967 Six Day War were un-modified from their original American-built configuration. Pictured is a M48A2C tank in IDF service during the Six Day War. The very cramped American-designed M1 cupola proved very unpopular with Israeli tank commanders. *Patton Museum*

Following the 1967 Six Day War, the IDF set about replacing the American-designed M1 cupola on its M48 tank series with a new low-profile cupola named the *"Urdan"* after the Israeli company that designed and built it. Pictured is an *Urdan* cupola on an IDF *Magach 3* tank. *Chris Hughes*

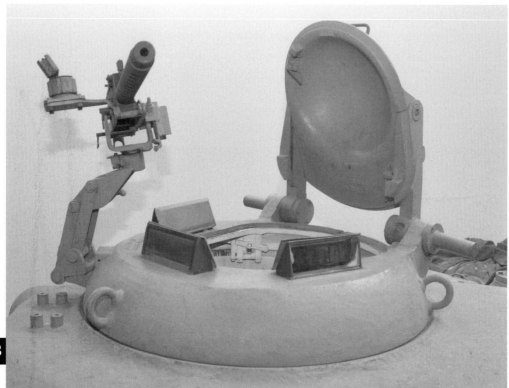

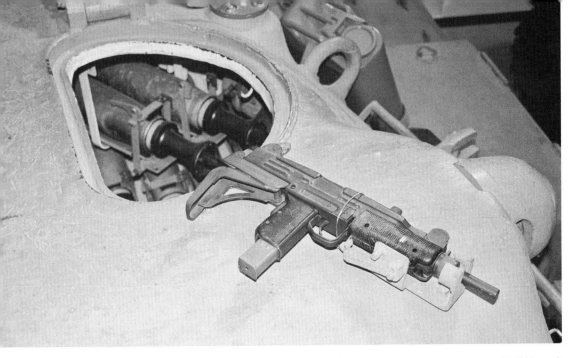

This close-up picture shows the bracket mounted in front of the loader's hatch on an IDF *Magach* 3 for storing an Israeli-designed and built Uzi submachine gun. Tank crews forced to abandon their vehicles for whatever reason in combat need to have some form of self- protection. *Michael Green*

An overhead view of the Israeli-designed and built *Urdan* cupola on a *Magach* 3 tank. Directly in front of the cupola is the armored cover for the gunner's sighting periscope. The *Urdan* cupola is very similar to the Chrysler-designed low profile cupola that appeared on the original M48 tank. *Chris Hughes*

This picture shows a formation of IDF *Magach* 3 tanks on patrol in the Sinai Peninsula sometime before the 1973 Yom Kippur War. Despite having been upgraded with a 105mm main gun and a diesel engine, the vehicles shown still have the unsatisfactory original American-designed and built M1 cupola. *IDF*

Taken during the 1973 Yom Kippur War is this picture of a *Magach* 3 that had not yet been upgraded with an *Urdan* cupola. For logistical reasons the IDF chose to keep its inventory of M48 and M60 series tanks fighting the Egyptian Army while its fleet of British-designed and built Centurion tanks fought on the Syrian Front. *IDF*

When IDF *Magach* 3s tried to rescue Israeli troops cut-off by advancing enemy units during the 1973 Yom Kippur War, they took heavy losses from Egyptian infantrymen armed with rocket-propelled grenade launchers (RPGs) and *Sagger* antitank guided missiles (ATGMs). This picture shows a knocked out IDF *Magach* 3 fitted with an *Urdan* cupola. *Egyptian Embassy*

Forming part of the collection of the Israeli Armored Corps Memorial Site and Museum is this *Magach* 3 with an *Urdan* low-profile cupola. The .50 caliber (12.7mm) machine gun fitted directly over the vehicle's gun shield was intended to take out targets that did not merit a main gun round. *Vladimir Yakubov*

This picture was taken from the loader's side of an IDF *Magach* 3 looking towards the tank commander's position. In the lower left foreground of the photograph is the recoil guard. Above the recoil guard is the rangefinder. The black handgrip is the tank commander's power control assembly, which affords him a capability identical to that of the gunner. *Michael Green*

Visible from the tank commander's position on an IDF *Magach* 3 is the gunner's seat. Directly in front of the gunner's seat are the black hand grips of his main gun control handles, nicknamed the "Cadillacs" by American tank crews because they were designed and built by the Cadillac Gage Company. *Michael Green*

Looking forward from the loader's seat on an IDF *Magach* 3 is part of the main gun ammunition storage arrangement, in this case loaded with armor piercing rounds. On the interior turret wall just above the stored main gun rounds is a large white-painted metal bin that contains the ammunition supply for the tank's 7.62mm coaxial machine gun. *Michael Green*

This picture shows a Taiwanese Army CM11 tank, officially nicknamed the "Brave Tiger," which consists of an M48A2 turret equipped with a 105mm main gun and an advanced fire-control system. The upgraded turret, which has an *Urdan* cupola, is mounted on the chassis of an M60A3 tank. *Chun-Lun Hsu*

A Taiwan Army CM11 tank drives past a reviewing stand during a public demonstration. It retains the rear turret bustle rack that first appeared on the M48A1 tank. Notice the wind senor stalk on the rear of the vehicle's turret. *Chun-Lun Hsu*

This picture taken during an open house day for the public at a Taiwan Army base shows a CM12 tank, which features the turret from the CM11 Brave Tiger mounted on the upgraded chassis of an M48A1 tank fitted with a diesel engine. Notice the M60 type headlight cluster and brush guards. *Mai Sen*

The Taiwanese Army did not have the engine compartment enlarged on their CM12 tanks for the fitting of a new diesel engine. Rather, they found an engine that would fit within the existing dimensions of the tank's rear hull with its flat engine deck and angled rear hull plate. *Mai Sen*

Chapter Eight:

Variants and Accessories

The U.S. Army is a firm believer in using existing tanks or components from those tanks in service as the basis for the specialized vehicles that are often developed to support the tanks. These include as part of the Patton tank family of vehicles: tank recovery vehicles, armored vehicle launched bridges, flamethrower tanks, self-propelled artillery, antiaircraft tanks, and combat

On display at the German Army Tank Museum is an American designed and built M88 armored recovery vehicle (ARV). The M88 was powered by a 12-cylinder, supercharged, fuel injected, air-cooled gasoline engine, designated the AVSI-1790-6A, which provided it with 980 horsepower. *Frank Schulz*

engineer vehicles. There were also accessories developed for the Patton tank series that included mine clearing equipment as well as flotation devices and fording equipment.

Armored Recovery Vehicles

An important vehicle to keep tanks up and moving on the battlefield is the armored recovery vehicle (ARV). The first ARV in U.S. Army service was based on a modified M3 medium tank chassis and designated the M31 tank recovery vehicle. It was replaced by an M4 series medium tank ARV designated the M32 series, which would remain in service with the American military after the Second World War.

Following the outbreak of the Korean War, a need was identified for a new more powerful ARV that could handle the heavier M46 series tanks. It was decided to use the second-generation M4A3 tank as the basis upon which a new and improved ARV could be built. The first of these new ARVs, designated the M74 tank recovery vehicle, rolled off the production line in February 1954. At

The gasoline engine in the M88 ARV was coupled to an Allison Cross-drive XT-1410-2A transmission, which provided it with three gears forward and one in reverse. The vehicle had a top sustained speed on a level road of 30 mph and a cruising range of approximately 200 miles. *Michael Green*

This picture taken during the Vietnam War of a U.S. Army M88 ARV shows various external details of the vehicle's upper hull, including the "A" frame type hoisting boom in its stowed position. The four man crew of the M88 ARV consisted of the vehicle commander, driver, mechanic, and rigger. *Patton Museum*

the same time, it was clear to many that the M74 was only a stopgap ARV as a new ARV based on a contemporary postwar medium tank was needed.

At first, it was envisioned that a modified M48 series tank could be employed as an ARV. This idea was eventually rejected and it was decided to design a new dedicated ARV using as many M48 series tank components as possible. That new ARV was standardized in February 1959 as the full tracked medium recovery vehicle M88. Over 1,000 of the gasoline-powered M88s ARVs came off the assembly lines between 1960 and 1964. The M88 weighed in at 110,000lbs (50mt) and had a cruising range of about 200 miles (322km).

Reflecting the fact that U.S. Army decided to switch to diesel-powered tanks

The erected "A" frame type hoisting boom of an M88 ARV appears in this picture. The boom was attached to the vehicle by a trunnion lever mounted on each side of the forward armored crew cab roof. The trunnion levers extended downward and connected to the hydraulic boom cylinders anchored to the vehicle's hull in the crew compartment. *TACOM*

The M88A1 is powered by an air-cooled, 12-cylinder Continental AVDS-1790-2DR diesel engine coupled to an Allison XT-1410-4 transmission. An external spotting feature for the M88A1 ARV is the M239 smoke grenade launchers pods mounted on either side of the vehicle's front hull, which are not seen on the M88 ARV. *TACOM*

in the 1960s, a need to re-engine the M88 ARV was obvious. However, it took until 1975 before the first diesel engine powered M88 ARV drove off the conversion line. The new version of the M88 series was designated the M88A1 ARV and widely exported to many of the countries that had the M48 and M60 series tanks in service.

Adequate for towing the M48 and M60 series tanks, the M88A1 ARV met its match with the M1 tank series that began appearing in U.S. Army service in 1981. It often took two M88A1s ARVs to tow a single disabled M1 series tank. In 1991, a decision was made to convert some 100 of the U.S. Army's inventory of M88A1s ARVs to a more powerful version designated the heavy equipment recovery combat utility lift & evacuation M88A2 and officially nicknamed the "Hercules." The first of these new 140,000lbs (63mt) M88A2s ARVs entered into U.S. Army service in 1997. The Marine Corps also took a small number of

Awaiting its next mission is a U.S. Marine Corps M88A2 ARV, officially nicknamed the "Hercules." This up-armored vehicles weighs 70 tons and is powered by an air-cooled, 12-cylinder Continental AVDS 1790-8CR engine producing over 1,000 horsepower. Notice the armored side shirts not seen on its predecessors. *Michael Green*

M88A2s ARVs into use. The vehicle has been exported to friendly foreign countries that have the M1 tank series in service, such as Egypt and Saudi Arabia.

Armored Vehicle Launched Bridges

During the Second World War it was the British Army who pioneered the use of the armored vehicle launched bridge (AVLB). The U.S. Army did not develop an interest in fielding an AVLB until 1949 when it was decided to use the chassis of an M46 series tank to mount a 63 foot span aluminum scissor bridge that could be launched and recovered with a hydraulically operated system. By the time the testing process proved the M46 tank based AVLB was a workable concept, the vehicle itself had been rendered obsolete. The U.S. Army quickly switched over to using the chassis of the M48 tank to see if the AVLB concept could be made to work again, which it did. When the green light was finally given to place the U.S. Army's first AVLB into service, the chassis of the M48A2C tank was selected because it was the most modern version of the M48 series tank

A U.S. Marine Corps M88A2 ARV is pictured towing a stuck M983 heavy expanded mobility tactical truck during a convoy in Afghanistan. The crew on this latest version of the M88 series ARVs is now down to three men, the vehicle commander, driver and mechanic. *U.S. Marine Corps*

then in service. Eventually, in the late 1970s, the entire inventory of M48A2C-based AVLBs was brought up to the M48A5 standard with diesel engines. All of these M48A5 based AVLBs were retired from American military service by the 1990s.

Flamethrower tanks

A small number of M4 series tanks were modified to serve as mechanized flamethrowers for use by the U.S. Army and U.S. Marine Corps during the Second World War. Primarily employed in the Pacific theater of operations against Japanese defensive positions, they proved extremely useful. Despite their proven usefulness in combat the U.S. Army decided in 1948 that it saw no future need for a dedicated flamethrower tank. The U.S. Marine Corps on the other hand was convinced by its Second World War experience with flamethrower tanks that they could still play an important role in certain combat environments. The Marine Corps would use some of its M4 series tanks modified into flamethrower tanks during the Korean War. They would remain in U.S. Marine Corps service until 1959.

To satisfy the Marine Corps desire for a more modern flamethrower tank, the U.S. Army Chemical Corps began work on a project in the late 1940s to mount a main armament flamethrower in the T42 medium tank. As that program came to an end and the turret of that never-fielded tank was mounted on the modified chassis of an M46 tank to produce the M47 tank, the Chemical Corps switched its effort over to produce a flamethrower version of the M47 tank, which was designated the T66. Only a single pilot example of the T66 was ever completed as the M47 tank was already being replaced in Marine Corps service with the M48 series tank.

The Chemical Corps quickly began to modify its main armament flamethrower, now designated as the M7-6, to fit inside a very early-production M48 tank with the Chrysler cupola. With the M7-6 mounted in the M48 tank the pilot vehicle was designated as the T67. The designation M7-6 referred to the fuel and pressure unit M7 and the flame gun M6.

A series of tests conducted by the Marine Corps soon confirmed that the pilot T67 flamethrower tank met all of it needs and a total of 74 vehicles based on the M48A1 tank entered into service as the flamethrower tank M67. The Marine

Taking part in a beach training exercise is a U.S. Marine Corps M88A2 ARV. Notice the fording equipment mounted on the rear of the vehicle's engine compartment. The M88A2 ARV is equipped with a Nuclear, Biological, and Chemical (NBC) protection system as well as a smoke screen generator. *U.S. Marine Corps*

The U.S. Army's first postwar attempt to come with an armored vehicle launched bridge (AVLB) was based on the chassis of the M46 series tank as shown here. The prototype vehicle pictured is emplacing a 63-foot span aluminum (with steel reinforcement) scissoring-type bridge, which was launched and retrieved hydraulically. *TACOM*

Corps crews that served on the M67 and its successor versions nicknamed them "flames" or "Zippos" after the well-known cigarette lighter.

As the M48A2 tank replaced the M48A1 tank in U.S. Army service, an improved main armament flamethrower designated the M7A1-6 went into the M48A2 tank, creating the flamethrower tank M67A1. Thirty-five units of the M67A1 were built between 1955 and 1956 for the U.S. Army, which used them for a limited period of time.

When the Marine Corps began replacing its M48A1 tanks with the modernized M48A3 tank it found the funding to have seventy-three of its M67 flamethrower tanks brought up to the M48A3 configuration, which resulted in the new designation as the full tracked combat tank, flame thrower M67A2. It

On display at the German Army Tank Museum is this AVLB based on the chassis of an M48A2 tank. The large flat metal plate just above the front of the tank's hull provides the base upon which the portable bridge rests as it is deployed into position *Frank Schulz*

was this version of the flamethrower tank that saw service with the Marine Corps during the Vietnam War. The M67A2 weighed in at approximately 107,000lbs (48mt).

Self-propelled artillery

The U.S. Army had quickly grasped the importance of having highly mobile, self-propelled artillery accompany its tanks into battle during the Second World War. Using the hull components of the M3 and M4 series medium tanks, the U.S. Army fielded the M7 series of 105mm howitzer motor carriages, nicknamed the "Priest" by the British military. A total of 3,490 M7s were completed. The U.S. Army also fielded during the latter part of the Second World War the 155mm gun motor carriage M12 based on the M3 series of medium tanks. Using M4 series medium tank components the U.S. Army also fielded the 155mm gun

A rear view of an AVLB, based on the chassis of an M48A5 tank, which is evident from the box-like M60 dry type air cleaner mounted on the right fender of the vehicle shown. Visible at the rear of the stowed bridge are the large hinge and associated cables that allow the two bridge sections to unfold and deploy. *Michael Green*

motor carriage M40 as well as the 8-inch howitzer motor carriage M43.

In the years after the Second World War it was decided to use as many components as possible from the next generation of medium tank to field a lightly armored chassis able to interchangeably mount a 155mm gun or 8-inch (203mm) howitzer. The end result of this project was the 155mm self-propelled gun M53 and the 8-inch self-propelled howitzer M55 fielded in 1956. Both vehicles borrowed the road wheels and the 23-inch (58cm) wide tracks found on the M46 and M47 series tanks. The engines and transmissions of the M53 and M55 were also the same as used in the M46 through M47 series tanks.

The U.S. Army took only the M55 configuration into service. The Marine Corps adopted the M53 and M55, both lasting in service all the way through the early stages of the Vietnam War. The M53 weighed in at about 100,000lbs (45mt) and the M55 roughly 98,000lbs (44mt).

On display at a Taiwan Army open house day is this AVLB, based on the chassis of an M48A5 tank. The 63-foot span aluminum scissoring type-bridge is seen at its full deployed height before it is either deployed or retracted. The bridge sits upon a large I-beam arrangement that is visible over the rear engine deck of the vehicle shown. *Mai Sen*

Antiaircraft tanks

During the Second World War the U.S. Army depended on a variety of antiaircraft guns mounted on armored halftracks for the protection of its frontline mechanized ground elements. Following the Second World War the U.S. Army fielded the full-tracked M19 self-propelled 40mm gun based on the chassis of the M24 light tank and the M42 self-propelled 40mm gun based on the chassis of the M41 light tank. All of these weapon systems depended on optical sights for target acquisition and tracking.

In the early 1970s, the U.S. Army identified a need for a turret-mounted,

A California Army National Guard AVLB based on the chassis of an M48A5 tank is shown in the process of recovering its 63-foot span aluminum scissoring type-bridge. The controls for launching and retrieving the bridge were located in front of the driver's seat. *Michael Green*

radar-guided mobile antiaircraft gun system that could defend its frontline mechanized ground elements from attack by low-flying enemy helicopter gunships and ground attack aircraft. The new antiaircraft gun system would have to fit on the chassis of a government-supplied M48A5 tank. The U.S. Army referred to this proposed antiaircraft tank as the Division Air Defense (DIVAD) gun system.

The eventual winner of the contract to provide the U.S. Army with the DIVAD was Ford Aerospace, which came up with a large two-man turret that mounted two 40mm Bofors L/70 guns. In an attempt to keep costs down, Ford Aerospace used a modified F-16 fighter radar system to guide the guns when firing. Sadly, testing of the Ford Aerospace DIVAD, designated the XM247 and officially nicknamed the "Sergeant York," in honor of the First World War Medal of Honor recipient U.S. Army Sergeant Alvin C. York, revealed it as an extremely unreliable and trouble-prone weapon system. Plagued by ever increasing cost overruns, the then-Secretary of Defense Caspar Weinberger canceled the program in 1985 after roughly fifty vehicles had been completed.

A side view of a U.S. Marine Corps flame thrower tank M67 that was based on a modified M48A1 tank. Since the vehicle had no cannon, the loader's position was done away with and replaced with a large container of napalm, bringing the crew down to three men. The M6 flame gun was housed in a metal shroud intended to mimic the appearance of the standard 90mm main gun. *Patton Museum*

Engineering Vehicles

The British Army fielded tank-based vehicles modified for use by the Royal Engineers during the Second World War. These specialized vehicles proved very successful during the latter part of the conflict and interested the U.S. Army enough to begin modifying a few of its M4 series tanks for use by its combat engineers. The war would end before any of the American tank-based engineering vehicles would see active service. In the years following the Second World War the U.S. Army attempted to field a specialized engineering vehicle based first on the M46 tank series and then later the M47 tank. Neither project was ever fielded. Eventually, the U.S. Army would field the full tracked combat engineer vehicle M728, based on the chassis of the M60 tank series.

Pictured is a U.S. Marine Corps full-tracked combat tank, flame thrower M67A2, which was based on a modified M48A3 tank chassis. Notice the flattened brush guards on the front hull-mounted headlight clusters to allow for more depression of the flame tube. *Marine Corps*

What the U.S. Army did field for use during the Second World War were the hydraulically operated M2 and M3 bulldozer kits that could be mounted on the M4 series tanks. These bulldozer kits were used by both regular armored units and combat engineer units.

In the immediate postwar period, the U.S. Army modified the M3 bulldozer kit to fit on the M47 tank. For use on the M48 and M48A1 tanks, the U.S. Army came up with a kit designated tank mounting bulldozer M8. For the M48A2 and M48A2C tanks there was an improved version designated as tank mounting bulldozer M8A1. The M48A3 and M48A5 tanks employed the tank mounting bulldozer M8A3. All of these M8 series bulldozer blades were raised and lowered hydraulically. Since the bulldozers blades fitted on M48 series tank blocked the vehicle's front headlights when in the raised travel position, the headlights were often elevated.

Mine Clearing Vehicles

From the Second World War and on, the most difficult project undertaken by the U.S. Army was the development of successful tank-mounted mine clearance equipment. During the Second World War the U.S. Army experimented with large roller type mine-exploder devices with mixed results. This line of development was picked up in the postwar years with the same desultory results and nothing being fielded. Everything was looked at, including the fitting of a large mine excavating plow on the front of an M47 tank, which did not work out as hoped.

The Vietnam War and the large number of mines encountered by American AFVs pushed the U.S. Army Mobility Equipment Research and Development Center to eventually deploy to South Vietnam twenty-seven experimental expendable mine clearing roller kits to be mounted on M48A3 tanks. This effort was

Spewing napalm from its M6 flame gun somewhere in South Vietnam is this U.S. Marine Corps full-tracked combat tank, flame thrower M67A2. In theory, the M6 flame gun had a maximum range of 200 yards. However, accuracy dropped off at 100 yards even under ideal conditions. *Patton Museum*

A U.S. Marine Corps full-tracked combat tank, flame thrower M67A2 is pictured in South Vietnam as refugees from the fighting stream past. Very evident in this picture is the shorter and stubbier appearance of the metal shroud for the M6 flame gun that identifies it as an M67A2 and not a standard M48A3 tank. *Patton Museum*

performed under the Expedited, Non-Standard, Urgent Requirement for Equipment (ENSURE) and as the mine clearing roller kit was the 202nd item under this program, it therefore was referred to as ENSURE 202. Sadly, the American tankers in the field were not impressed by the usefulness of the ENSURE 202 and the device was not retained in service after the Vietnam War ended.

The German Army has an inventory of twenty-four specialized mine clearing flail vehicles built on the heavily modified chassis of the M48 series tank. The official nickname for the vehicle is *Keiler*, which translates into English as the "Wild Boar."

Floatation Devices and Fording Equipment

Combat experience gained during the Second World War convinced the U.S. Army and Marine Corps of the need to get tanks ashore as quickly as possible during amphibious operations or river crossing operations. Following this train of thought the U.S. Army developed the T15 flotation device for the M47 tank, and the T44 flotation device for the M48 series tanks.

Belonging to a private collector is this ex-U.S. Marine Corps 155mm self-propelled gun M53. The vehicle's turret was limited to a traverse of 30 degrees right and left with a hydraulic system or when manually operated. The vehicle had a crew of six men. *Michael Green*

Both flotation devices consisted of four large float assemblies strapped together, which surrounded the tanks in question. The float located directly behind the tanks contained two rudders that would allow the vehicle and attached floats to be steered. Attached propellers were driven by the tank's drive sprockets. Both flotation devices could be jettisoned by the tank crews once their vehicle's had safely made it ashore. Because of their bulk both flotation devices proved impractical for use and work on this concept was eventually ended.

The introduction of the M48 tank prompted the U.S. Army to come up with a deep water fording kit that would take advantage of the vehicle's air-cooled engine ability to operate fully submerged in water. Modified versions of the deep water fording kit were also developed for subsequent version of the M48 series tanks as well as the M60 series tanks. Despite the misleading name of the deep water fording kit. It was primarily intended to allow a vehicle to transit from a naval landing craft through a surf line and then onto a beach.

The main external feature of the deep water fording kit on the M48A2 tank and follow-on versions was a tall vertical stack attached to the rear of the tank's engine compartment to vent exhaust gases. The engine air intake came through the tank's crew compartment, which would have been waterproofed and fitted with a bilge pump to take care of any leakage problems. The exhaust stack on the tank was attached to a lanyard so the tank commander could pull it down once safely ashore, to save it from damage, reduce the vehicle's silhouette and permit 360 degree turret traverse.

The 8-inch self-propelled howitzer M55 pictured belongs to the U.S. Army Ordnance Museum collection. Armor protection on the M55 was limited to 13mm, except for the vehicle's front hull plate that was 25mm thick. The M55 had storage space for 10 rounds of 8-inch ammunition onboard. *Christopher Vallier*

Clearly visible in this picture is the large rear hull spade of a 155mm self-propelled gun M53. The spade was lowered prior to action and helped to stabilize the gun when firing and absorbed some of the weapon's recoil. *Michael Green*

A Ford Aerospace Division Air Defense (DIVAD) gun system, also known as the M247 Sergeant York, is shown posed for a publicity picture. The two 40mm guns in the 360 degree rotating turret had a rate of fire per gun of 600 rounds per minute and could reach out and strike targets at about seven miles (12.5km). *TACOM*

The M247 Sergeant York was intended to accompany the U.S. Army M1 series tanks and the M2/M3 series of infantry and cavalry fighting vehicles on the battlefield and protect them from both enemy attack helicopters and ground attack aircraft. However, the mobility of the M48A5 tank chassis was severely lacking compared to the vehicles it was supposed to protect. *TACOM*

Among the many reasons that doomed the poorly conceived M247 Sergeant York to cancellation was the fact that the Soviet Army had fielded Tactical Air-To-Surface Missiles (TASM) on its helicopter gunships that could be launched from outside the maximum range of the Sergeant York's twin 40mm guns. *TACOM*

The U.S. Army engineer armored vehicle T39 was based on the chassis of the M46 series tanks. It was armed with the 165mm gun T156, which fired a 30 pound high explosive plastic (HEP) projectile to destroy heavily protected targets, such as bunkers, at short range. In British military parlance the round would be referred to as high explosive squash head (HESH) projectile. *Patton Museum*

A rear view of the engineer armored vehicle T39 based on the chassis of the M46 series tanks. Visible on the rear engine deck of the vehicle is the housing for a 20-ton winch. There was storage space inside the vehicle for 26 main gun rounds. *Patton Museum*

Mounted on the front of this M47 tank is the tank mounted, bulldozer M3. This was the replacement for the M1 and M2 series bulldozer employed on the M4 series tanks during the Second World War. *Patton Museum*

A factory photograph taken at the Detroit Arsenal shows an M48A1 tank with the tank-mounting, bulldozer M8, shown in its lowered position. What is interesting in this picture is the fact that the headlight arrays have been moved to the vehicle's front fenders. *TACOM*

This picture taken during a large-scale training exercise in South Korea shows a South Korean Army M48A5PI equipped with a tank mounted bulldozer M8A3. The bulldozer blade is shown in its raised travel position. Notice the raised headlight arrays on this tank. *TACOM*

An interesting mine clearance device experimentally fitted to the M47 tank was the Peter Pan light mine excavating plow seen here. In theory, the V-shaped plow dug into the soil as it was pushed forward and in the process funneled any mines in its path to the sides of the vehicles where they could later be disarmed. *TACOM*

The first postwar roller type mine exploder tested by the U.S. Army was nicknamed the "Larruping Lou" and is seen here mounted on the front of an M48 tank. A boom assembly attached to the front of the tank's hull guided twelve large steel discs to hopefully detonate any mines in the vehicle's path. *Patton Museum*

Pictured mounted to the front of a U.S. Army M48A3 tank during the Vietnam War is the ENSURE 202 expandable mine clearing roller. The 10,000 pound device was made up of 12 modified tank road wheels divided into two separate six-wheel roller assemblies that were each 47 ½ inches wide. *TACOM*

Mounted on the front of this Israeli *Magach* 3 is an Israeli copy of a Soviet Army tank mounted mine roller, which is on display at the Israeli Armored Corps Memorial Site and Museum. Visible on the vehicle hull and turret are attachment points for Blazer explosive reactive armor tiles. *Vladimir Yakubov*

Pictured is an M47 tank afloat with the aid of the T15 flotation device during tests by the Detroit Arsenal on June 3, 1953. The T15 flotation device with attached tank was 41 feet long and had a width of 24 feet. *TACOM*

Coming off a U.S. Navy landing craft during the Vietnam War is a U.S. Marine Corps M48A3 tank. The vehicle is fitted with the deep water fording kit. Clearly visible at the rear of the engine compartment is the tall vertical fording exhaust stack. *Patton Museum*